Passive Network Synthesis

Advances in Design and Control

SIAM's Advances in Design and Control series consists of texts and monographs dealing with all areas of design and control and their applications. Topics of interest include shape optimization, multidisciplinary design, trajectory optimization, feedback, and optimal control. The series focuses on the mathematical and computational aspects of engineering design and control that are usable in a wide variety of scientific and engineering disciplines.

Series Volumes

Morelli, A. and Smith, M. C., *Passive Network Synthesis: An Approach to Classification*
Özbay, Hitay, Gümüşsoy, Suat, Kashima, Kenji, and Yamamoto, Yutaka, *Frequency Domain Techniques for $\mathcal{H}_\infty$ Control of Distributed Parameter Systems*
Khalil, Hassan K., *High-Gain Observers in Nonlinear Feedback Control*
Bauso, Dario, *Game Theory with Engineering Applications*
Corless, M., King, C., Shorten, R., and Wirth, F., *AIMD Dynamics and Distributed Resource Allocation*
Walker, Shawn W., *The Shapes of Things: A Practical Guide to Differential Geometry and the Shape Derivative*
Michiels, Wim and Niculescu, Silviu-Iulian, *Stability, Control, and Computation for Time-Delay Systems: An Eigenvalue-Based Approach, Second Edition*
Narang-Siddarth, Anshu and Valasek, John, *Nonlinear Time Scale Systems in Standard and Nonstandard Forms: Analysis and Control*
Bekiaris-Liberis, Nikolaos and Krstic, Miroslav, *Nonlinear Control Under Nonconstant Delays*
Osmolovskii, Nikolai P. and Maurer, Helmut, *Applications to Regular and Bang-Bang Control: Second-Order Necessary and Sufficient Optimality Conditions in Calculus of Variations and Optimal Control*
Biegler, Lorenz T., Campbell, Stephen L., and Mehrmann, Volker, eds., *Control and Optimization with Differential-Algebraic Constraints*
Delfour, M. C. and Zolésio, J.-P., *Shapes and Geometries: Metrics, Analysis, Differential Calculus, and Optimization, Second Edition*
Hovakimyan, Naira and Cao, Chengyu, *$\mathcal{L}_1$ Adaptive Control Theory: Guaranteed Robustness with Fast Adaptation*
Speyer, Jason L. and Jacobson, David H., *Primer on Optimal Control Theory*
Betts, John T., *Practical Methods for Optimal Control and Estimation Using Nonlinear Programming, Second Edition*
Shima, Tal and Rasmussen, Steven, eds., *UAV Cooperative Decision and Control: Challenges and Practical Approaches*
Speyer, Jason L. and Chung, Walter H., *Stochastic Processes, Estimation, and Control*
Krstic, Miroslav and Smyshlyaev, Andrey, *Boundary Control of PDEs: A Course on Backstepping Designs*
Ito, Kazufumi and Kunisch, Karl, *Lagrange Multiplier Approach to Variational Problems and Applications*
Xue, Dingyü, Chen, YangQuan, and Atherton, Derek P., *Linear Feedback Control: Analysis and Design with MATLAB*
Hanson, Floyd B., *Applied Stochastic Processes and Control for Jump-Diffusions: Modeling, Analysis, and Computation*
Michiels, Wim and Niculescu, Silviu-Iulian, *Stability and Stabilization of Time-Delay Systems: An Eigenvalue-Based Approach*
Ioannou, Petros and Fidan, Barış, *Adaptive Control Tutorial*
Bhaya, Amit and Kaszkurewicz, Eugenius, *Control Perspectives on Numerical Algorithms and Matrix Problems*
Robinett III, Rush D., Wilson, David G., Eisler, G. Richard, and Hurtado, John E., *Applied Dynamic Programming for Optimization of Dynamical Systems*
Huang, J., *Nonlinear Output Regulation: Theory and Applications*
Haslinger, J. and Mäkinen, R. A. E., *Introduction to Shape Optimization: Theory, Approximation, and Computation*
Antoulas, Athanasios C., *Approximation of Large-Scale Dynamical Systems*
Gunzburger, Max D., *Perspectives in Flow Control and Optimization*
Delfour, M. C. and Zolésio, J.-P., *Shapes and Geometries: Analysis, Differential Calculus, and Optimization*
Betts, John T., *Practical Methods for Optimal Control Using Nonlinear Programming*
El Ghaoui, Laurent and Niculescu, Silviu-Iulian, eds., *Advances in Linear Matrix Inequality Methods in Control*
Helton, J. William and James, Matthew R., *Extending $\mathcal{H}^\infty$ Control to Nonlinear Systems: Control of Nonlinear Systems to Achieve Performance Objectives*

Passive Network Synthesis
An Approach to Classification

A. Morelli
University of Cambridge
Cambridge, United Kingdom

M. C. Smith
University of Cambridge
Cambridge, United Kingdom

Society for Industrial and Applied Mathematics
Philadelphia

10 9 8 7 6 5 4 3 2 1

Publications Director	Kivmars H. Bowling
Executive Editor	Elizabeth Greenspan
Developmental Editor	Gina Rinelli Harris
Managing Editor	Kelly Thomas
Production Editor	David Riegelhaupt
Copy Editor	Susan Fleshman
Production Manager	Donna Witzleben
Production Coordinator	Cally A. Shrader
Compositor	Cheryl Hufnagle
Graphic Designer	Doug Smock

Library of Congress Cataloging-in-Publication Data

Names: Morelli, A. (Alessandro), 1987- author. | Smith, Malcolm C. (Malcolm Clive), 1957- author.

Title: Passive network synthesis : an approach to classification / A. Morelli, M. C. Smith, University of Cambridge, Cambridge, United Kingdom.

Description: Philadelphia : Society for Industrial and Applied Mathematics, [2019] | Series: Advances in design and control ; 33 | Includes bibliographical references and index.

Identifiers: LCCN 2019005292 (print) | LCCN 2019017504 (ebook) | ISBN 9781611975826 | ISBN 9781611975819 (print)

Subjects: LCSH: System analysis.

Classification: LCC T57.6 (ebook) | LCC T57.6 .M64 2019 (print) | DDC 003--dc23

LC record available at *https://lccn.loc.gov/2019005292*

Contents

Chapter 1
Introduction

The central goal of network synthesis is to devise a network which realizes a prescribed behavior. Implicit in this task is the characterization of the behaviors that are in principle realizable with certain specified components and those that are not. In passive network synthesis the specified components are the standard (passive) electrical elements such as the resistor, capacitor, and inductor or their mechanical equivalents. In contrast to synthesis, network analysis studies the behavior of a *given* circuit which obeys the relevant component laws and the interconnection rules, e.g., Kirchhoff's law. Thus, synthesis implicitly contains analysis and has a higher level of complexity.

The first major result of electrical network synthesis, and arguably the birth of network synthesis, is the 1924 publication of the reactance theorem of Foster [25]. The theorem gives a necessary and sufficient condition that specifies the impedance function of an arbitrary network comprising inductors, capacitors, and transformers only. An explicit construction is given for a network realizing any impedance of the prescribed class. With this theorem Foster established network synthesis as an exemplar of an engineering theory that sought to describe rigorously the entities which are possible to realize physically in a given domain.

Network synthesis developed rapidly in the following decades. By the 1960s a corpus of results had been established which is now considered classical. Foster's theorem was generalized to two-element-kind networks by Cauer [10]. Brune [8] introduced the important class of positive-real functions, which are those functions that are capable of being realized as a two-terminal network with arbitrary (linear) passive components. Brune provided an explicit construction method for a general positive-real function which made use of resistors, inductors, capacitors, and transformers, as did a later one by Darlington [19]. A remarkable theorem of Bott and Duffin [7] showed that transformers were unnecessary.

Network synthesis flourished as an active research topic in the first half of the twentieth century due to the broad scope it offered and due to the practical motivation of developing useful results for analogue filter design. Research gradually petered out in the 1960s with the advent of integrated circuits. Nevertheless, the classical theory continued to exert considerable influence in related fields such as control and systems and is still felt strongly today. Linear systems theory, dissipativity, the parametrization of stabilizing controllers, conservation laws, systems on graphs, etc., all owe a debt to network synthesis.

In the last decade there has been a resurgence of interest in network synthesis motivated in part by the introduction of the *inerter*, a new fundamental element for mechanical control, and independently due to the advocacy of Kalman [51]. From both perspectives the classical theory remained incomplete in regard to transformerless synthesis. This is highlighted in the Bott–Duffin theorem. The construction is described in a single-page paper in the *Journal of Applied Physics*, yet the sharpness of the result is in stark contrast to the lack of intuitiveness of the proof. In his exposition of the construction [33, Chapter 10] Guillemin writes that it is difficult "to make the succession of steps appear natural and meaningful except that they do ultimately lead to the desired result."

The most unusual aspect of the Bott–Duffin proof is that the number of energy storage elements (inductors and capacitors) in the realization appears excessive compared to the degree of the impedance function. There were a number of independent attempts to simplify the construction [23, 62, 68] with marginal success. Foster conceded in [27] that no further simplifications seemed likely by similar methods and issued a challenge "to derive a much better method of synthesis." No further progress was achieved for half a century.

Recent work has established that the Bott–Duffin construction is the simplest possible among series-parallel networks for certain biquadratic impedance functions [42], and that the known simplifications cannot be improved in general [39], apart from some newly discovered circuits of equal complexity. At one level Foster's challenge is answered in the negative. But at another, much still remains to be better understood, since a great many positive-real functions can be realized in a much simpler manner.

The present state of knowledge brings an earlier approach to RLC synthesis into the foreground again: the enumerative approach. An example of this is the Master's thesis of Ladenheim [56], a student of Foster at the Polytechnic Institute of Brooklyn. Ladenheim determines the class of all essentially distinct two-terminal electrical networks comprising at most two reactive elements (capacitors or inductors) and at most three resistors—now known as the "Ladenheim catalogue." Until recent years Ladenheim's dissertation appears to have been virtually unknown. A single citation in [28] independently led to two publications: one by Jiang and Smith [47] and the other by Kalman [51].

In the Ladenheim catalogue the impedance for each network is computed, and the inverse process is performed; i.e., given the impedance, an expression for each element of the network is stated. However, there are no derivations in Ladenheim's work and his catalogue leaves several important questions unanswered, the most important being the absence of conditions on the impedance coefficients which ensure positivity of the network parameters. Furthermore, Ladenheim only considers what will be referred to here as *generic* networks, yet that concept is only implicit in the thesis.

The purpose of this monograph is to provide an outline of the main (classical and modern) contributions to the field of passive network synthesis and to present new research into the field on the enumerative approach and the classification of networks of restricted complexity. In particular, we provide a new analysis and classification of the Ladenheim catalogue, building on recent work, to obtain an improved understanding of the structure and realization power of the class within the biquadratic positive-real functions. The motivation is twofold. With the introduction of the inerter, recent applications of network synthesis in the mechanical domain have highlighted the need for a better understanding of the most "economical" way to realize a given passive impedance. The Ladenheim class, comprising at most two reactive elements, is the simplest class that is nontrivial and is therefore ideal for an in-depth analysis. More broadly, the fundamental nature of the theorem of Bott and Duffin alone invites further study. Recent results have established that RLC realization is less structured than realization of linear dynamical systems, and this does suggest that Foster's challenge should be modified to a search for an improved general understanding of RLC synthesis.

1.1 ▪ Outline

Fundamental notions of two-terminal networks are reviewed in Chapter 2, and classical results from passive network synthesis are described. Modern developments in the field are summarized in Chapter 3, where relevant results on the classification of biquadratics are introduced. The inerter device, the mechanical-electrical analogy, and some applications of passive network synthesis to mechanical networks are then described in Chapter 4. One of the main contributions of this work is a complete analysis of the Ladenheim catalogue. The formal derivation of the catalogue is described in Chapter 5, where the group action and equivalence relations, which will be our main tools for the classification of networks, are introduced. For each equivalence class, the set of impedances that can be realized is derived in explicit form as a semi-algebraic set. Realizability conditions, expressed in terms of necessary and sufficient conditions, are given in Chapter 6 for each equivalence class. Based on such derivations, the main results are formalized in Chapter 7, where a new notion of generic network is introduced and observations are made on the complete set of equivalences, on the smallest generating set for the class, and on Kalman's latest work. The class of six-element networks with four resistors is also analyzed in Chapter 7. Finally, conclusions are drawn in Chapter 8. A series of appendices which are useful in the study of the catalogue are provided, including the realization theorems for all the five-element networks in the catalogue.

1.2 ▪ Acknowledgments

The Ladenheim catalogue was a central part of discussions at the four workshops on mathematical aspects of network synthesis initiated by Uwe Helmke which were held alternately in Würzburg and Cambridge from 2010 to 2016, and it is from these roots that the present monograph has emerged. The monograph owes a special debt to Uwe Helmke and two of the participants of the early workshops, Rudolf Kalman and Jan Willems, all three of whom have sadly passed away since 2010.

The authors are most grateful to Tim Hughes, Jason Jiang, Rodolphe Sepulchre, and Paolo Rapisarda for their helpful comments on the manuscript. The first author acknowledges the support of The MathWorks for its funding of The MathWorks studentship in Engineering at the University of Cambridge.

Chapter 2

Classical results of network synthesis

A short survey is undertaken in this chapter of general background and classical results of passive network synthesis to provide a broader context for the subsequent analysis. A survey of more recent work and applications can be found in Chapters 3 and 4. For further details and material on the results discussed in this chapter see [1, 12, 33, 87, 102].

2.1 ▪ Preliminaries of electrical networks

This monograph is concerned with linear passive electrical networks, with a particular focus on one-ports. A one-port network (also know as a two-terminal network), as shown in Figure 2.1, has two external terminals (nodes, vertices), 1 and $1'$. The voltage v between the terminals of the port and the current i entering one terminal and leaving from the other is taken with the sign convention of Figure 2.1. One-ports may be connected to other one-ports, for instance, another element, a network, or a driving source (voltage or current generating source). The *driving-point impedance* of N is defined by $Z(s) = \hat{v}(s)/\hat{i}(s)$, where ^ denotes the Laplace transform, and $Y(s) = Z^{-1}(s)$ is the driving-point admittance.

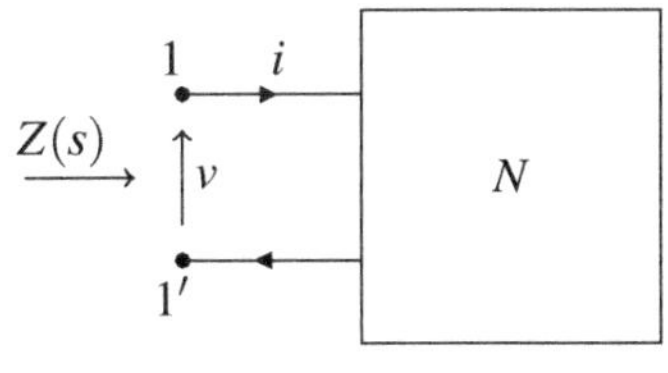

Figure 2.1. *Electrical network N with two external terminals 1 and $1'$, terminal voltage v, terminal current i, and driving-point impedance $Z(s)$.*

Networks will comprise finite interconnections satisfying Kirchoff's laws that contain resistors, inductors, and capacitors and will be referred to as RLC networks. See Figure 2.2 for the standard symbols of the network elements. From time to time we shall have occasion to refer to more general one-ports and multiports containing ideal

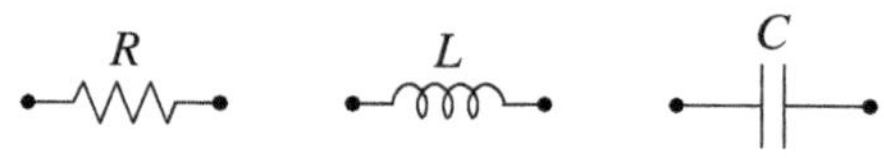

Figure 2.2. *The three fundamental two-terminal electrical elements (one-ports): resistor, inductor, and capacitor.*

transformers or coupled coils, though they do not form part of this study. The inductor and capacitor are termed *reactive elements* and have impedance sL and $1/(Cs)$, respectively, where $L > 0$ and $C > 0$ denote inductance and capacitance. The resistor has impedance R, where $R > 0$ denotes the resistance.

Associated to each network is an undirected connected graph in which each edge (branch) corresponds to a network element and two special vertices (nodes) are identified, i.e., the external terminals. We restrict our attention to *planar* networks, whose graph can be embedded in a plane in such a way that no two edges intersect. A corollary of a famous theorem by Kuratowski establishes that graphs with fewer than five nodes or nine edges must be planar [74, Theorem 3-17]. In this study we deal with networks with fewer than nine edges; therefore, we do not need to worry about nonplanarity.

A graph has a *dual* if and only if it is planar [74, Theorem 3-15]. Given a planar graph G, the vertices of its dual G' each correspond to a face of G, while faces of G' each correspond to a vertex of G. Two vertices in G' are connected by an edge if the corresponding faces in G have an edge in common. A graph and its dual always have the same number of edges, since there is a one-to-one correspondence between them [74, Section 3-3], while the number of vertices need not be the same. The dual of a planar graph is not necessarily unique, in the sense that the same graph can have nonisomorphic dual graphs (which can stem from distinct planar embeddings of the same graph). However, if G_1 and G_2 are dual graphs of the same planar graph, then G_1 and G_2 are 2-isomorphic [74, Theorem 3-18] (see also Section 5.1 and [98, 97] regarding 2-isomorphism).

In the case of two-terminal electrical networks, the first step to obtain the dual of a network N is to consider a (voltage or current) driving source connected to the two external terminals: this step is needed in order to preserve the port in the duality process, since it would otherwise be replaced by a short circuit in the dual network. We then consider the graph G associated to the network and construct its dual G' according to the method outlined above. The dual network N' can then be obtained from G' by populating each edge in G' with the dual of the corresponding element in N: inductors are replaced by capacitors of equal value and vice versa, and resistors are replaced by resistors of reciprocal value [74, Section 6-6]. The voltage source is replaced by a current source and vice versa, which allows us to identify the two external terminals in the dual network [32, Chapter 10.9]. An example of a simple three-element RLC network and its dual is given in Figure 2.3, while Figure 2.4 illustrates the procedure which leads to the graph dual of the given network.

Series-parallel (SP) networks are two-terminal networks which can be constructed inductively by combining other SP networks either in series or parallel, with single-element networks being SP. Networks that are not SP are termed *bridge networks*. Networks obtained through a series (parallel) connection of two SP networks are termed *essentially series* (*essentially parallel*) [70]. *Simple series-parallel* (SSP) networks are SP networks which can be constructed inductively by combining a single element in series or in parallel with an SSP network, with single-element networks being SSP. Similar definitions can be given for the graphs associated to the networks.

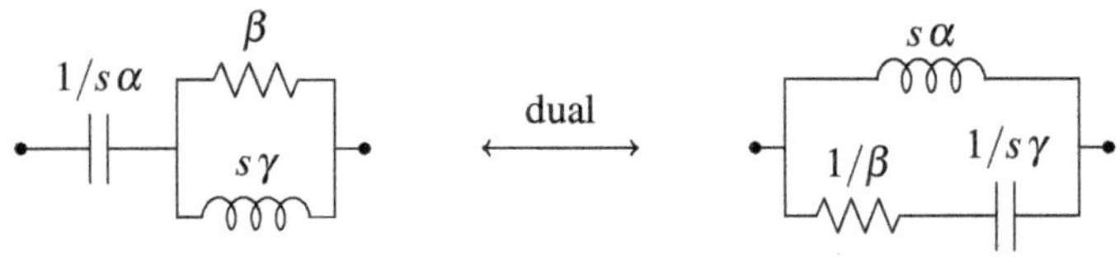

Figure 2.3. *Two-terminal electrical network and its dual, with element values α, β, $\gamma > 0$. The impedance of each element in the network is the reciprocal of the corresponding element in the dual network.*

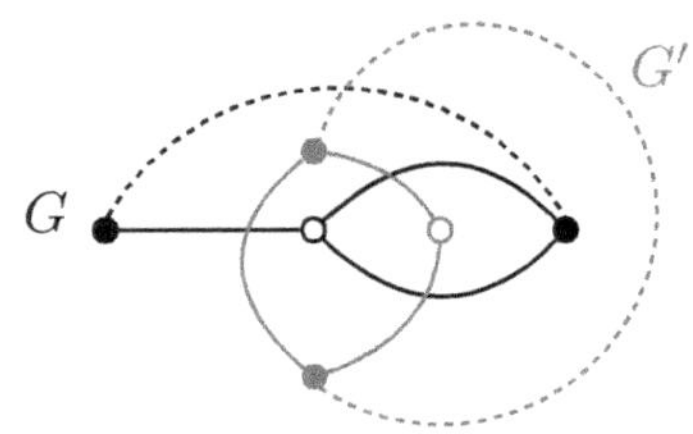

Figure 2.4. *Graphical illustration of the procedure to obtain the dual of a given graph. The original graph G is drawn in black, with the dashed line indicating the additional edge corresponding to a voltage or current source connecting the driving-point terminals of G. The same holds for the dual graph G', which is drawn in grey. Note that G' has a vertex for every face of G and that there is a one-to-one correspondence between edges of G and edges of G'.*

2.2 ▪ Foster and Cauer canonical forms

Network synthesis in its modern sense originated in the famous reactance theorem of Foster [25]. The result completely characterizes the one-ports that can be built with reactive elements only. The proof of the theorem is based on the solution of an analogous dynamical problem of the small oscillations of a mechanical system given by E. J. Routh [72]. The theorem takes the following form.

Theorem 2.1. *The most general driving-point impedance obtainable in a passive network without resistors (LC network) takes the form*

$$Z(s) = k\left[\frac{(s^2+\omega_1^2)(s^2+\omega_3^2)\dots(s^2+\omega_{2n\pm 1}^2)}{s(s^2+\omega_2^2)(s^2+\omega_4^2)\dots(s^2+\omega_{2n}^2)}\right]^{\pm 1}, \tag{2.1}$$

where $k \geq 0$, $0 < \omega_1 < \omega_2 \dots$, and $n \geq 0$. Any such impedance may be physically realized in the form of Figure 2.5(a) *or Figure* 2.5(b) *through a partial fraction expansion of $Z(s)$ or $Z^{-1}(s)$ in the form*

$$k_0 s + \frac{k_\infty}{s} + \sum_{r=1}^{m} \frac{s k_r}{s^2 + p_r^2}, \tag{2.2}$$

where $k_r \geq 0$, $p_r > 0$ (distinct), and $m \geq 1$.

Two alternative realizations of purely reactive networks were introduced soon afterwards by Cauer [10] and arise from continued fraction expansions of $Z(s)$—see Figures 2.5(c) and 2.5(d). Such expansions may be developed from $Z(s)$ by an alternating sequence of operations of the form "extract a pole" and "invert" where the simple pole extracted is at $s = 0$ or $s = \infty$. The continued fraction for Figure 2.5(c) takes the form

$$Z(s) = sL_1 + \cfrac{1}{sC_1 + \cfrac{1}{sL_2 + \cfrac{1}{sC_2 + \dots}}} ,$$

while that for Figure 2.5(d) takes the form

$$Z(s) = \frac{1}{sC_1} + \cfrac{1}{\frac{1}{sL_1} + \cfrac{1}{\frac{1}{sC_2} + \cfrac{1}{\frac{1}{sL_2} + \dots}}} .$$

Theorem 2.1 shows that the impedance of a purely reactive network has by necessity only simple poles and zeros which alternate on the imaginary axis, with $s = 0$ and $s = \infty$ always a simple pole or zero. Cauer [10] showed that a similar situation applies to other two-element-kind networks, for which canonical networks analogous to those of Figure 2.5 can be obtained. By analogy to the LC case, these networks are also termed Foster and Cauer canonical forms.

Theorem 2.2. *The most general driving-point impedance obtainable in a passive network which contains only resistors and capacitors (RC network) takes the form*

$$Z(s) = k\frac{(s+\lambda_1)(s+\lambda_3)\dots(s+\lambda_{2n\pm1})}{(s+\lambda_0)(s+\lambda_2)\dots(s+\lambda_{2n})}, \tag{2.3}$$

where $k \geq 0$, $0 \leq \lambda_0 < \lambda_1 < \lambda_2 \dots$, and $n \geq 0$. Any such impedance may be physically realized in the form of Figure 2.5(a) *or Figure* 2.5(b) *(with inductors replaced by resistors) through a partial fraction expansion of $Z(s)$ or $Y(s)/s$, or in the form of Figure* 2.5(c) *or Figure* 2.5(d) *through a continued fraction expansion.*

Theorem 2.3. *The most general driving-point impedance obtainable in a passive network which contains only resistors and inductors (RL network) takes the form*

$$Z(s) = k\frac{(s+\lambda_0)(s+\lambda_2)\dots(s+\lambda_{2n})}{(s+\lambda_1)(s+\lambda_3)\dots(s+\lambda_{2n\pm1})}, \tag{2.4}$$

where $k \geq 0$, $0 \leq \lambda_0 < \lambda_1 < \lambda_2 \dots$, and $n \geq 0$. Any such impedance may be physically realized in the form of Figure 2.5(a) *or Figure* 2.5(b) *(with capacitors replaced by resistors) through a partial fraction expansion of $Z(s)/s$ or $Y(s)$, or in the form of Figure* 2.5(c) *or Figure* 2.5(d) *through a continued fraction expansion.*

It is worth pointing out that the general form of the impedance in (2.1), (2.3), and (2.4) is not altered if the networks contain transformers, and yet transformers are not needed in the canonical forms.

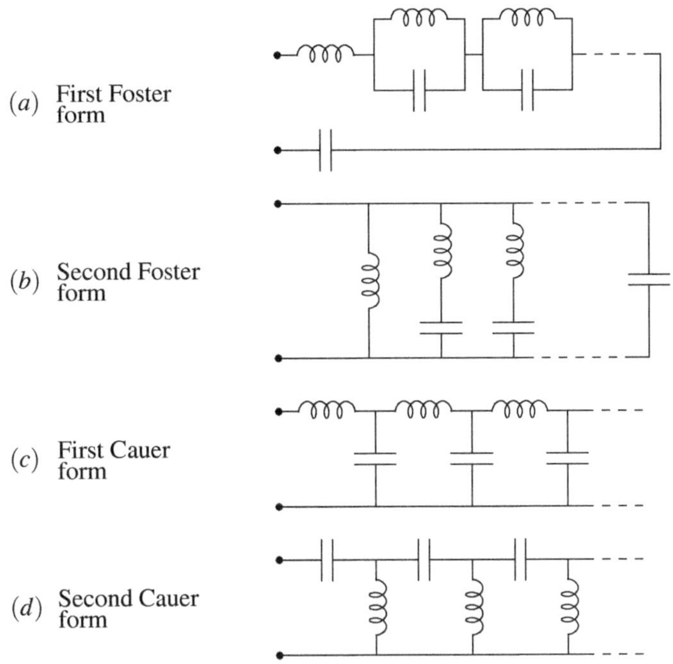

Figure 2.5. *Foster and Cauer canonical forms for two-element-kind networks (LC case). For the RC case, replace L by R; for the RL case, replace C by R.*

2.3 ▪ Positive-real functions and passivity

A significant further step in the development of passive network synthesis was the paper of Brune [8]. His first important contribution was to introduce and give a complete characterization of the class of positive-real functions and further to show that the driving-point impedance of any passive one-port network must be positive-real.

Brune's derivation in [8] that the driving-point impedance is positive-real if and only if the network is passive is based on a physical argument which we now outline. Given the one-port passive network N in Figure 2.1 with driving-point impedance $Z(s)$, consider a voltage impulse applied to its terminals. For a passive network the resulting transient current will be bounded, and hence it is necessary that all the zeros in $Z(s)$ have nonpositive real part. By a similar argument, considering a current impulse at the port, it is necessary that all the poles in $Z(s)$ have nonpositive real part. As we will see next, much stronger properties hold for the driving-point impedance of a passive network.

Consider now an applied voltage $v(t) = 0$ for $t < 0$ and $v(t) = e^{\gamma t}\cos(\omega t)$ for $t \geq 0$, with $\gamma > 0$. We can calculate that the forced response takes the form

$$i(t) = \frac{e^{\gamma t}}{|Z(\gamma + j\omega)|}\cos(\omega t - \theta)\,, \quad \text{where } \theta = \arg Z(\gamma + j\omega)\,,$$

and that this dominates any free response (transient) terms. Neglecting the transient terms, the energy delivered to the network can be computed as

$$W = \int_0^T v(t)i(t)\,dt = \frac{1}{4\,|\gamma + j\omega| \cdot |Z(\gamma + j\omega)|} \Big\{ e^{2\gamma\tau} \cos(2\omega\tau - \theta - \chi) - \cos(\theta + \chi) + \frac{\cos\theta}{\cos\chi}(e^{2\gamma\tau} - 1) \Big\},$$

where $\chi = \arg(\gamma + j\omega)$. If τ is large enough, the terms in $e^{2\gamma\tau}$ will be dominant, and we can conclude that $\cos\theta / \cos\chi \geq 1$ necessarily, where both numerator and denominator must be positive, since $\gamma > 0$. If $s = \gamma + j\omega$, this can be rewritten in the form of the following two conditions, which represent the base for Brune's definition of positive-real functions from which all other necessary conditions follow:

$$\operatorname{Re}(Z(s)) \geq 0 \quad \text{for} \quad \operatorname{Re}(s) \geq 0\,, \tag{2.5}$$

$$|\arg Z(s)| \leq |\arg s| \quad \text{for} \quad |\arg s| \leq \pi/2\,. \tag{2.6}$$

We note that (2.6) can be interpreted as a contraction property of such functions: the phase of the function is always smaller in absolute value than the phase of its argument. Condition (2.6) clearly implies (2.5), and Brune proved the remarkable result that the converse also holds using Pick's theorem, a generalization of Schwarz's lemma [8, Theorem VII]. The two conditions can therefore be considered as equivalent definitions of positive-realness for a real-rational function $Z(s)$.

We now state two more commonly used definitions of positive-real functions, which are entirely equivalent to the definitions given in (2.5) and (2.6) [1, Section 2.7], [8, Theorem V]. We note that conditions similar to those in Theorem 2.5 can be given in terms of the zeros of $Z(s)$ [8, Theorem V, Corollary 1].

Definition 2.4. *A rational function $Z(s)$ is defined to be positive-real if the following hold:*

1. *$Z(s)$ is real for real s;*
2. *$Z(s)$ is analytic for $\operatorname{Re}(s) > 0$;*
3. *$\operatorname{Re}(Z(s)) \geq 0$ for $\operatorname{Re}(s) > 0$.*

Theorem 2.5. *A rational function $Z(s)$ is defined to be positive-real if the following hold:*

1. *$Z(s)$ is real for real s;*
2. *$Z(s)$ is analytic for $\operatorname{Re}(s) > 0$;*
3. *poles on the imaginary axis are simple and have positive real residues;*
4. *$\operatorname{Re}(Z(j\omega)) \geq 0$ for all ω.*

We further state some useful properties of positive-real functions which appear in Brune's work and which all follow from the definition of positive-realness.

Theorem 2.6. *If $Z(s)$ is a positive-real function, then the following properties hold:*

1. *$1/Z(s)$ is positive-real;*

2. *$Z(1/s)$ is positive-real;*

3. *the degree of the numerator and denominator of $Z(s)$ can differ by at most one;*

4. *the real part of $Z(s)$ in the right half-plane attains its minimum value on the imaginary axis;*

5. *any poles on the extended imaginary axis can be extracted as in a partial fraction expansion, with the terms extracted and the remainder necessarily positive-real;*

6. $|\arg Z(s)| \leq |\arg s|$ *for* $|\arg s| \leq \pi/2$.

We conclude by stating a more formal definition of passivity [61, Definition 2.5], which can be shown to be equivalent to positive-realness [61, Theorem 4.3]. The definition formalizes the notion that the total energy delivered to the network up to time τ is nonnegative, meaning that no energy can be delivered to the environment.

Definition 2.7. *A one-port network with driving-point voltage $v(t)$ and driving-point current $i(t)$ is passive if*

$$\int_{-\infty}^{\tau} v(t)i(t)\, dt \geq 0$$

for all τ and for all compatible pairs $v(\cdot)$, $i(\cdot)$ which are square integrable on $(-\infty, \tau]$.

A similar definition with the lower limit replaced by t_0 is given in [1], with the additional assumption that the network is storing no energy at time t_0. A proof of the equivalence with positive-realness is given in [1, Theorem 2.7.3].

2.4 ▪ The Foster preamble and Brune cycle

The second important contribution of Brune [8] was to formulate a procedure to find a network that realizes an arbitrary positive-real function. The procedure begins with a sequence of steps known as the *Foster preamble*. This involves the removal of any imaginary axis poles or zeros from $Z(s)$ and the reduction of its minimum real part to zero. For example, if $Z(s)$ has a pole at $s = \infty$, then we can write

$$Z(s) = sL + Z_1(s),$$

where $L > 0$ and $Z_1(s)$ is positive-real with no pole at $s = \infty$. In network terms this corresponds to the removal of a series inductor as shown in the top left figure in Table 2.1. Poles at the origin or on the imaginary axis are dealt with in a similar way. Zeros are similarly extracted by the corresponding operations on $Y(s) = 1/Z(s)$. At any point a constant equal to the minimum value of the real part of the function can be subtracted from $Z(s)$ or $Y(s) = 1/Z(s)$. A summary of the different operations which can be performed on either $Z(s)$ or $1/Z(s)$ is given in Table 2.1, in terms of network representations. The process is not unique.

If the process described above does not succeed in completely realizing the function, then a positive-real function $Z(s)$ remains which has no poles or zeros on the

Table 2.1. *Summary of the possible removal operations on an impedance function $Z(s)$ or an admittance function $Y(s)$. In each case the remainder of this extraction ($Z_1(s)$ or $Y_1(s)$) is positive-real by Theorem 2.6. The removal of zeros on the extended imaginary axis corresponds to the removal of poles from the reciprocal function. See [87] for a textbook explanation.*

	from impedance function $Z(s)$	from admittance function $Y(s)$
Removal of a pole at infinity	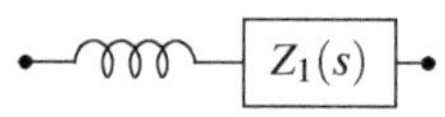	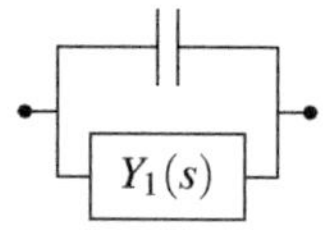
Removal of a pole at zero	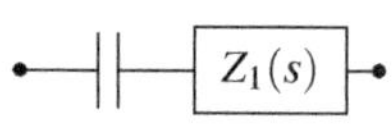	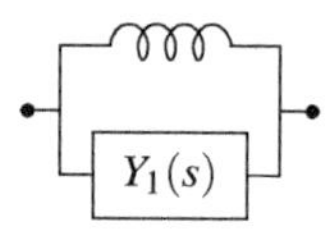
Removal of a pair of imaginary-axis poles	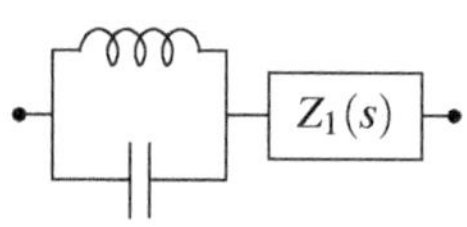	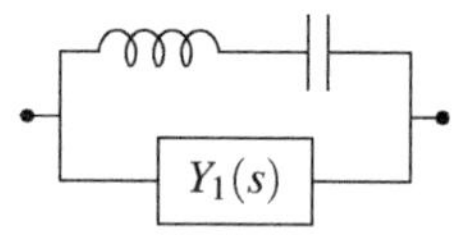
Removal of a constant	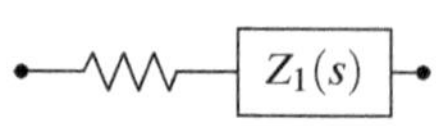	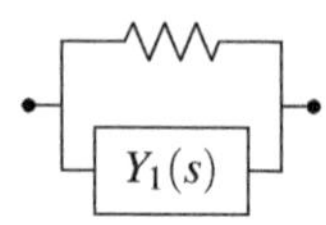

extended imaginary axis and with the real part of $Z(j\omega)$ equal to zero at one or more finite, nonzero frequencies. Such functions are termed *minimum functions*. There then follows an operation known as a *Brune cycle* which extracts three inductors in a Y-configuration together with a capacitor. No matter how this extraction is performed, one of the inductors is required to have a negative inductance—which is not realizable passively. Brune's decisive step is to show that the Y-configuration can always be replaced by a pair of coupled coils of positive inductance and with a unity coupling coefficient (Figure 2.6). Such a transformer is in principle realizable passively, though difficult in practice.

We conclude by highlighting the main result from Section 2.3 and the present section in the following theorem, known as Brune's theorem.

Theorem 2.8.

1. *The driving-point impedance $Z(s)$ of any linear passive one-port network is positive-real.*
2. *If $Z(s)$ is positive-real, then it is realizable by a network containing resistors, inductors, capacitors, and ideal coupled coils.*

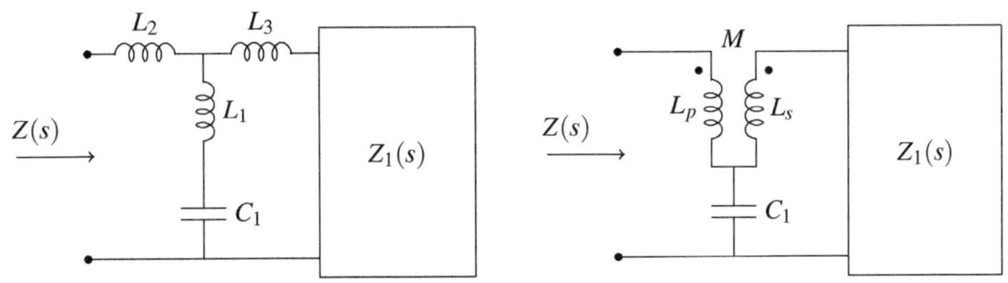

Figure 2.6. *Replacement of the Y-configuration of inductors (where $L_1 > 0$, and one of L_2 and L_3 is negative) with coupled coils in a Brune cycle, with L_p, $L_s > 0$ and $M^2/(L_pL_s) = 1$ (unity coupling coefficient).*

2.5 ▪ The Bott–Duffin construction and its simplifications

The remarkable fact that coupled coils or ideal transformers can be dispensed with in the realization of positive-real functions was shown by Bott and Duffin [7]. Their construction begins in the same way as Brune's method, leading to a minimum function by the Foster preamble. The key step of the method is a replacement for the Brune cycle. This makes use of the Richards transformation [69], which states that, for any positive-real function $Z(s)$ and any $k > 0$,

$$R(s) = \frac{kZ(s) - sZ(k)}{kZ(k) - sZ(s)}$$

is positive-real of degree no greater than $Z(s)$. Now suppose that $Z(s)$ is a minimum function with $Z(j\omega_1) = j\omega_1 X_1$ where $\omega_1 > 0$ and assume that $X_1 > 0$ (otherwise the argument is applied to $Z^{-1}(s)$). Then we can find a $k > 0$ so that $R(s)$ has a zero at $s = j\omega_1$, by choosing $Z(k)/k = X_1$. We now write

$$Z(s) = \frac{kZ(k)R(s) + Z(k)s}{k + sR(s)}$$

$$= \frac{1}{\dfrac{1}{Z(k)R(s)} + \dfrac{s}{kZ(k)}} + \frac{1}{\dfrac{k}{Z(k)s} + \dfrac{R(s)}{Z(k)}} \tag{2.7}$$

$$= \frac{1}{\dfrac{1}{\dfrac{kZ(k)}{s} + \dfrac{Z(k)}{R(s)}} + \dfrac{1}{Z(k)R(s) + \dfrac{sZ(k)}{k}}} \tag{2.8}$$

and note that (2.7) and (2.8) correspond to the circuit diagrams of Figure 2.7(a)–(b). We can then write

$$\frac{1}{R(s)} = \frac{\gamma s}{s^2 + \omega_1^2} + \frac{1}{R_1(s)}$$

for some $\gamma > 0$ and $R_1(s)$ being positive-real of strictly lower degree than $R(s)$, which allows series or parallel resonant circuits to be extracted to obtain the networks shown

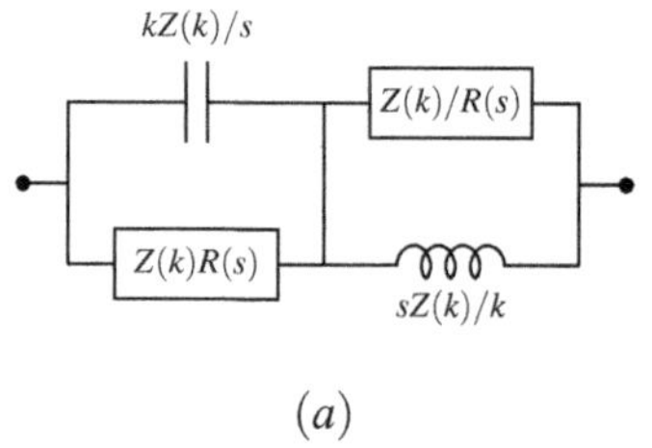

Figure 2.7. *Networks realizing the inverse of the Richards transformation.*

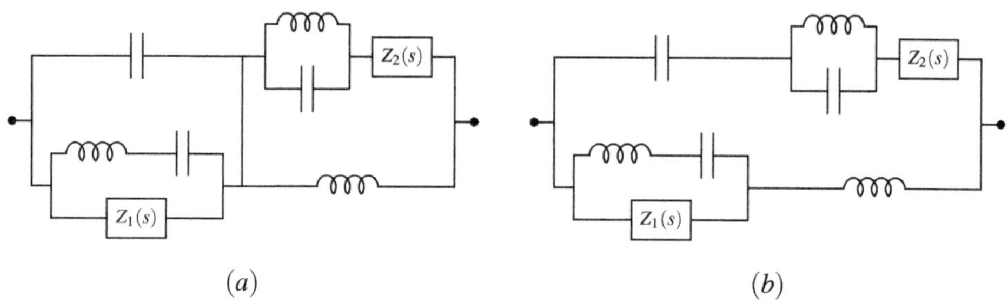

Figure 2.8. *Bott–Duffin cycle for the minimum function $Z(s)$ in which $Z_1(s)$, $Z_2(s)$ have lower degree than $Z(s)$ in the case $X_1 > 0$.*

in Figure 2.8. The Bott–Duffin method continues as necessary on the reduced degree impedances until resistors are obtained.

Given the above construction method, we can now state the Bott–Duffin theorem.

Theorem 2.9. *Any positive-real function can be realized as the driving-point impedance of a network containing resistors, inductors, and capacitors only.*

It should be noted that the networks of Figure 2.8(a)–(b) contain six reactive elements for a degree reduction of two from $Z(s)$ to $Z_1(s)$ and $Z_2(s)$. If $Z(s)$ is a biquadratic minimum function, then $Z_1(s)$ and $Z_2(s)$ are resistors and six reactive elements in total are used in the network realization of $Z(s)$. This apparent extravagance prompted attempts to seek simpler realizations. Several authors independently found that the six reactive elements could be reduced to five if bridge networks were allowed [23, 62, 68]. The resulting network can be most easily derived by noting that the network in Figure 2.8(a) is a balanced bridge. Hence it must be entirely equivalent to the network of Figure 2.9. It turns out that by judicious choice of the additional inductance a Y-Δ transformation can be made which results in a pair of reactive elements being duplicated, and hence such a pair can be removed. The resulting network is shown in Figure 2.10. See [33] for a textbook explanation.

Despite the nonintuitive nature of these constructions, it has recently been shown that the Bott–Duffin construction is the simplest possible among SP networks for the biquadratic minimum function [42] and that the Pantell–Fialkow–Gerst–Reza simplification cannot be improved upon in the generic case [39]. It follows that "nonminimality" is intrinsic to the RLC realization of some driving-point impedances. This has

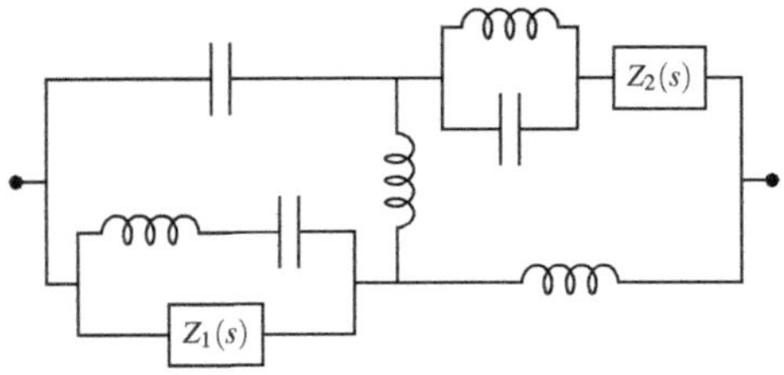

Figure 2.9. *Bott–Duffin network with additional inductor.*

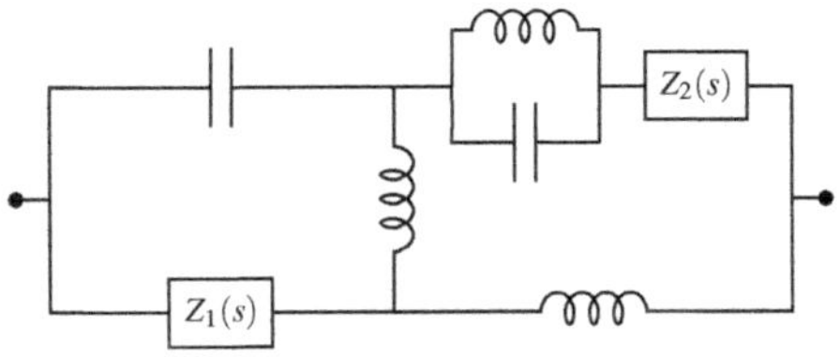

Figure 2.10. *Puntell–Fialkow–Gerst–Reza simplification of the Bott–Duffin cycle.*

prompted a fundamental treatment of this nonminimality using Willems' behavioral framework [38, 43]. See Section 3.4 for a discussion on the behavioral approach to passivity.

2.6 ▪ Darlington synthesis

Roughly midway between the appearance of the Brune and Bott–Duffin methods a completely different procedure to realize driving-point impedances was devised by Darlington [19]. Like Brune's approach, ideal transformers are an integral part. Remarkably, only one resistor is needed, no matter how complex the positive-real function. In the present context this serves to emphazise the extra freedom that is obtained when transformers are allowed. The Darlington theorem states the following.

Theorem 2.10. *Given a positive-real impedance, it can always be realized as the driving-point impedance of a lossless (i.e., comprising no resistive elements) two-port network terminated in a single resistance, as shown in Figure* 2.11.

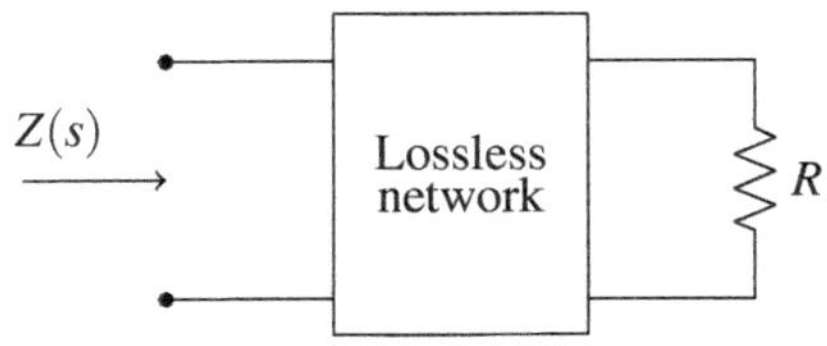

Figure 2.11. *Network structure for the Darlington method.*

We now provide an outline of Darlington's realization procedure, which proves the above result. We note that the resistance R which is extracted is often chosen to be 1Ω, though it may be set to any positive value.

Darlington's procedure made use of a lossless two-port containing inductors, capacitors, and transformers only. Such a two-port is "reciprocal," which means that the impedance matrix is symmetric [1]:

$$X(s) = \begin{pmatrix} x_{11}(s) & x_{12}(s) \\ x_{12}(s) & x_{22}(s) \end{pmatrix}. \tag{2.9}$$

Assuming that the two-port lossless network in Figure 2.11 has impedance matrix (2.9), the driving-point impedance $Z(s)$ is given by

$$Z(s) = x_{11} \frac{R^{-1}(x_{11}x_{22} - x_{12}^2)/x_{11} + 1}{R^{-1}x_{22} + 1}. \tag{2.10}$$

We now write $Z(s)$ as a ratio of polynomials in the form

$$Z(s) = \frac{m_1 + n_1}{m_2 + n_2}, \tag{2.11}$$

where m_1, m_2 are polynomials of even powers of s and n_1, n_2 are polynomials of odd powers of s. If we factor m_1 out of the numerator and n_2 out of the denominator, we get

$$Z(s) = \frac{m_1}{n_2} \frac{n_1/m_1 + 1}{m_2/n_2 + 1},$$

which, by comparison with (2.10), suggests the identification

$$x_{11} = \frac{m_1}{n_2}, \qquad x_{22} = R\frac{m_2}{n_2}, \qquad x_{12} = \sqrt{R}\,\frac{\sqrt{m_1 m_2 - n_1 n_2}}{n_2},$$

providing $m_1 m_2 - n_1 n_2$ is a perfect square. The above corresponds to case A in [33, Chapter 9.6]. The alternative case B is obtained by factoring n_1 out of the numerator and m_2 out of the denominator in (2.11), and leads to the same expressions for x_{11}, x_{22}, and x_{12}, with the letters n and m interchanged. It is possible to achieve a real-rational function for x_{12} in one of the two cases by allowing, as necessary, the introduction of cancelling factors between numerator and denominator in (2.11).

It can finally be shown that the set of transfer impedances x_{11}, x_{22}, and x_{12} which have been identified define an impedance matrix $X(s)$ which allows a realization as a lossless two-port comprising inductors, capacitors, and transformers only. This is achieved through an extension of Foster's synthesis method of Section 2.2 to the case of two-port networks, which was derived by Cauer [11]. In particular, a partial fraction expansion of $X(s)$, similar to (2.2), leads to an expression of the following type:

$$X(s) = sC_0 + \frac{1}{s}C_1 + \frac{s}{s^2 + p_2^2}C_2 + \frac{s}{s^2 + p_3^2}C_3 + \dots,$$

where $p_i > 0$ and C_0, C_1, etc., are nonnegative definite constant matrices. A typical term in this sum can always be realized in the form of a T-circuit, like the one shown in Figure 2.12, and connected in series to the other two-ports, as shown in Figure 2.13 (see [33, Chapter 7] for further details). Note that if $Z(s)$ has poles or zeros on the imaginary axis, then these may be extracted using the relevant steps in the Foster preamble (as outlined in Section 2.4) and included directly in the lossless two-port.

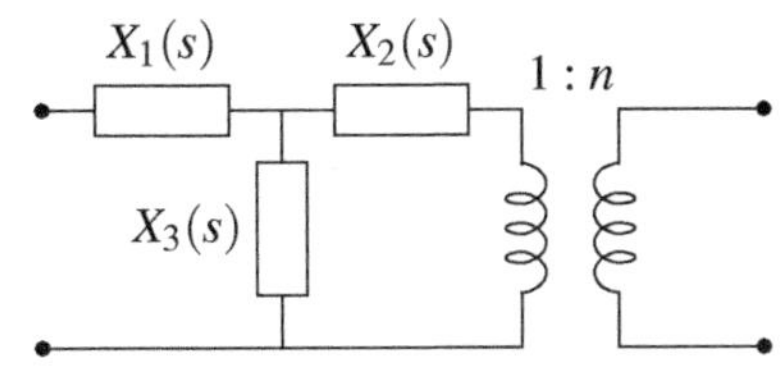

Figure 2.12. *Lossless two-port realization of a typical term in the Darlington synthesis, where $X_1(s)$, $X_2(s)$, and $X_3(s)$ are the impedances of an inductor, capacitor, or parallel LC circuit.*

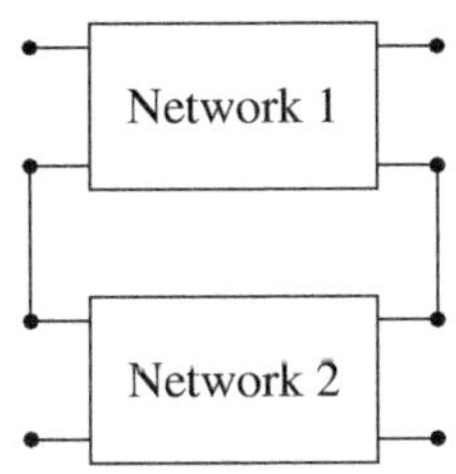

Figure 2.13. *Series connection of two two-port networks corresponding to the sum of two terms in the Darlington synthesis.*

2.7 ▪ Reactance extraction

A later approach to driving-point synthesis due to Youla and Tissi [103] is complementary to Darlington's. The framework is illustrated in Figure 2.14 and the approach is termed *reactance extraction*, in contrast to Darlington's approach of *resistance extraction*. Youla and Tissi's main result is the following.

Theorem 2.11. *Given a positive-real impedance, it can always be realized as the driving-point impedance of a nondynamic (i.e., comprising only resistors and ideal transformers) multiport network where all inductors and capacitors have been extracted, as shown in Figure* 2.14.

The approach of [103] established a connection with the state-space approach to linear dynamical systems which allowed matrix methods to be applied to the synthesis problem (see [1, 102] for textbook treatments). If there are p inductors and q capacitors which are extracted as in Figure 2.14 and if $\mathbf{i}_a = [i_1, \ldots, i_p]^T$, $\mathbf{i}_b = [i_{p+1}, \ldots, i_{p+q}]^T$ denote the vectors of Laplace-transformed port currents (and similarly for the corresponding port voltage vectors $\mathbf{v}_a$ and $\mathbf{v}_b$), then, under mild conditions, there exists a constant matrix M such that the multiport N_r is described by

$$\begin{pmatrix} v \\ \mathbf{v}_a \\ \mathbf{i}_b \end{pmatrix} = M \begin{pmatrix} i \\ \mathbf{i}_a \\ \mathbf{v}_b \end{pmatrix} = \begin{pmatrix} M_{11} & M_{12} \\ M_{21} & M_{22} \end{pmatrix} \begin{pmatrix} i \\ \mathbf{i}_a \\ \mathbf{v}_b \end{pmatrix}$$

with $JM = M^T J$, $J = \text{diag}(I_{p+1}, -I_q)$, and M partitioned so that M_{11} is the $(1,1)$

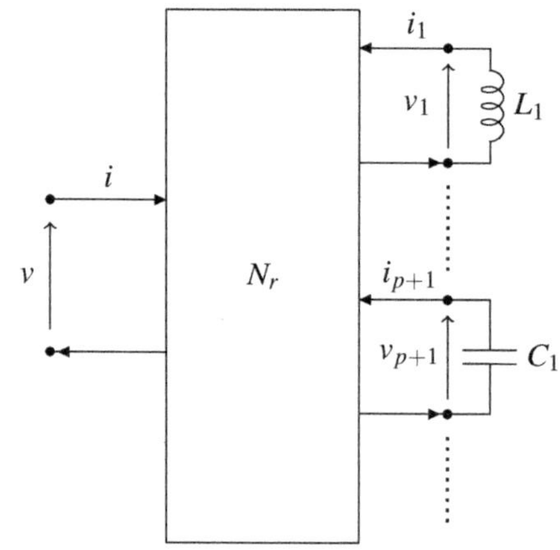

Figure 2.14. *Network structure for the method of reactance extraction, where the network N_r contains only resistors and transformers.*

element of M. Writing $\Lambda = \text{diag}(L_1, \ldots, L_p, C_1, \ldots, C_q)$ it follows that

$$Z(s) = M_{11} - M_{12}\left(sI + \Lambda^{-1}M_{22}\right)^{-1}\Lambda^{-1}M_{21},$$

which is an expression for the impedance in state-space form.

The contribution of [103] was to find all solutions of the multiport reciprocal synthesis problem using a minimum number of inductors and capacitors and making use of multiport transformers in the realization of N_r. The specific idea of extracting reactances in the form of Figure 2.14 had also been used earlier in [2], and subsequently in [67], to study the class of all biquadratic impedances that may be realized without transformers. This use relies on a condition known as *paramountcy* for the impedance matrix of a three-port to be realizable using resistors only. In particular, if M is a symmetric $n \times n$ matrix, with $n \leq 3$, then a necessary and sufficient condition for M to be realizable as the impedance matrix of an n-port network comprising resistors only is that M is paramount, namely each principal minor of the matrix is not less than the absolute value of any minor built from the same rows [13, 77]. We note that if $n > 3$, then this condition is only necessary. Also, if we allow transformers to be present in the realization, then a general necessary and sufficient condition for M to be realizable is simply that it is nonnegative definite (see, e.g., [1, Chapter 9]).

The results of [103] have been exploited recently in [41] to establish algebraic criteria for the number of inductors and capacitors present in a realization of an impedance function of a one-port network, as will be discussed in more detail in Section 3.3.

Chapter 3

Modern developments of network synthesis

Despite a golden period of advances starting in the 1920s and a wealth of elegant results, research in network synthesis slowly declined in the second half of the twentieth century, following the introduction of integrated circuits and the dwindling importance of analogue filters. Nevertheless, the basic results retained their fundamental importance, and the influence of circuit theory and network synthesis extended outside the electrical domain. Significant results in the systems and control community testify to an enduring importance and relevance of the subject: the Kalman–Yakubovich–Popov lemma (or "positive-real lemma") relating passivity to positive-realness, dissipativity theory [100], behavioral modeling [99], etc.

Many questions in network synthesis were, however, still unanswered and some results not fully understood, an example being the apparent nonminimality of the Bott–Duffin networks and its simplifications [39, 42]. Interest in efficient realizations of passive mechanical networks, following the invention of the inerter mechanical device [78], prompted a fresh look at these questions. In this chapter we review the most important results in relation to these modern developments of network synthesis, starting from the concept of regularity and Jiang and Smith's analysis of the class of five-element networks [47, 49]. (See also [40] for a survey of some of the results discussed in this chapter.)

3.1 ▪ Regular positive-real functions and the Ladenheim catalogue

The Master's thesis of Ladenheim [56] is probably the first example of systematic enumeration and analysis of all RLC networks within a restricted class. This is the class of all essentially distinct two-terminal RLC networks with at most five elements of which at most two are reactive. It follows that the driving-point impedance of such networks is at most biquadratic. There are 108 networks in the class, which we refer to as the "Ladenheim catalogue," of which a formal derivation is given in Chapter 5.

The approach of Jiang and Smith was to study the realization power of the relevant class of enumerated networks using the notion of a *regular* positive-real function.

Definition 3.1. *A positive-real function $Z(s)$ is defined to be regular if the smallest value of* $\mathrm{Re}(Z(j\omega))$ *or* $\mathrm{Re}(Z^{-1}(j\omega))$ *occurs at* $\omega = 0$ *or* $\omega = \infty$.

For biquadratic functions regularity implies that the Foster preamble succeeds in reducing the function to a resistor, which means that a realization is possible with at most two reactive elements and three resistors. It was shown in [47] that six such networks suffice to realize any regular biquadratic. A series of lemmas also showed that all but two of the 108 Ladenheim networks can realize only regular biquadratic impedances and that the remaining two networks are capable of realizing some but not all of the nonregular biquadratics. The Ladenheim class was thus shown to possess a *generating set* comprising eight circuits (see also Section 7.3). Reichert's theorem, which is discussed in more detail in Section 3.2, shows that additional resistors beyond three do not extend the class of functions that are realized by the class, which establishes that not all positive-real functions have an RLC realization with the total number of reactive elements being equal to the degree of the impedance.

The regularity concept has been further utilized to seek networks with more than two reactive elements that are capable of realizing nonregular biquadratics. Building on previous work of Vasiliu, five-element structures with three reactive elements were investigated in [47] and SP networks with six elements in [49]. The realizability region for all networks was characterized using a canonical form for biquadratics (which is reviewed in Section 3.1.2 below), and taken together, these networks were seen to be insufficient to cover the whole of the nonregular region for biquadratics.

An interesting survey by Kalman of the development of passive network synthesis from its early origins until the mid 1970s, when research on the topic gradually petered out, is given in [51]. Considerable attention is paid to Ladenheim's dissertation and the possibilities for such an enumeration approach to provide a better understanding of transformerless synthesis. Such an approach is further outlined in [59] (see also [81]) and in [52, 53], where the potential role of algebraic invariant theory as a "natural and effective tool for the network synthesis problem" is stressed.

The analysis in this monograph of the complete Ladenheim catalogue can be seen as a continuation of several lines of thinking on the problem. Although [47] identified a generating set for the catalogue, a detailed analysis of the realization power of all the networks was not undertaken. Recently Chen et al. [18] derived realizability conditions for the regular bridge networks of the Ladenheim class. This still left unknown the actual set of realizable impedance functions for many networks in the catalogue. Further, the multiplicity of solutions to the realization problem was not known for most networks. Also, some networks are known to be equivalent to others, but the full set of equivalences had not been determined. Further, the smallest generating set for the catalogue had not been clearly established. More broadly, the amount of structure in the class was a matter of conjecture. The analysis carried out in Chapters 5, 6, and 7 is intended to answer, or improve understanding on, all these questions.

3.1.1 ▪ Positive-real and regular biquadratics

In this section we will review some relevant results from [47] on the classification of biquadratic impedances. As mentioned above, the concept of regularity greatly facilitates the classification of impedances, and Lemmas 1–8 in [47] provide useful properties of regular functions. We restrict our attention to *biquadratic* impedances of the form

$$Z(s) = \frac{As^2 + Bs + C}{Ds^2 + Es + F}, \tag{3.1}$$

where $A, B, C, D, E, F \geq 0$. This function is positive-real if and only if

$$\sigma = BE - (\sqrt{AF} - \sqrt{CD})^2 \geq 0 \tag{3.2}$$

(see [17, Corollary 11]). We now look for conditions under which the biquadratic (3.1) is regular. The resultant of the numerator and denominator in (3.1), that is,

$$K = (AF - CD)^2 - (AE - BD)(BF - CE), \tag{3.3}$$

plays an important role in answering this question. Its sign determines whether the reactive elements in a realization of the biquadratic are of the same kind or of a different kind (see Section 3.3). If $K < 0$, the reactive elements are of the same kind and the numerator and denominator in (3.1) have real, distinct roots which interlace each other; by Lemma 3 in [47] the impedance is regular. If $K = 0$, the numerator and denominator have a root in common and the biquadratic function reduces to a bilinear function or a constant; by Lemma 7 in [47] the impedance is regular. Finally, if $K > 0$, the network will have one inductor and one capacitor and need not be regular. By Lemma 5 in [47] the biquadratic impedance (3.1) is regular if and only if at least one of the following four cases is satisfied:

$$\text{Case (1)} \quad AF - CD \geq 0 \text{ and } \lambda_1 \geq 0\,, \tag{3.4}$$
$$\text{Case (2)} \quad AF - CD \geq 0 \text{ and } \lambda_2 \geq 0\,, \tag{3.5}$$
$$\text{Case (3)} \quad AF - CD \leq 0 \text{ and } \lambda_3 \geq 0\,, \tag{3.6}$$
$$\text{Case (4)} \quad AF - CD \leq 0 \text{ and } \lambda_4 \geq 0\,, \tag{3.7}$$

where

$$\lambda_1 = E(BF - CE) - F(AF - CD)\,, \tag{3.8}$$
$$\lambda_2 = B(AE - BD) - A(AF - CD)\,, \tag{3.9}$$
$$\lambda_3 = D(AF - CD) - E(AE - BD)\,, \tag{3.10}$$
$$\lambda_4 = C(AF - CD) - B(BF - CE)\,. \tag{3.11}$$

3.1.2 ▪ A canonical form for biquadratics

The analysis of the five-element networks in the Ladenheim catalogue is aided by a *canonical form* for biquadratics [47, 67]. For the impedance $Z(s)$ two simple transformations can be defined:

1. multiplication by a constant α,
2. frequency scaling: $s \to \beta s$.

It is easily seen that these transformations correspond to the following scalings of the network parameters: $R \to \alpha R$, $L \to \alpha\beta L$, $C \to \beta C/\alpha$. With such transformations the biquadratic (3.1) with $A, \ldots, F > 0$ can always be reduced to the canonical form

$$Z_c(s) = \frac{s^2 + 2U\sqrt{W}s + W}{s^2 + (2V/\sqrt{W})s + 1/W} \quad (U, V, W > 0)\,, \tag{3.12}$$

where $\alpha = D/A$, $\beta = \sqrt[4]{CF/(AD)}$, $W = \sqrt{CD/(AF)}$, $U = B/(2\sqrt{AC})$, and $V = E/(2\sqrt{DF})$. We note that U corresponds to the damping ratio of the zeros of the biquadratic (3.1), and V is the damping ratio of the poles, while W is the ratio of the natural frequencies of zeros and poles.

The introduction of the canonical form reduces the number of coefficients from six in (3.1) to just three in (3.12), hence allowing an intuitive graphical interpretation of the realizable set for a given network. It is in fact possible, for a given value of W, to illustrate the set of values in the (U, V)-plane corresponding to real positive values of all inductances, capacitances, and resistances in the network. We will call such a set the *realizability region* of a network for a given W (see also Section 5.2.1). Figure 3.1 shows the regions in the (U, V)-plane which correspond to a regular biquadratic for $W \leq 1$, as also shown in [47]. For $K \leq 0$ the biquadratic is always regular, while for $K > 0$ the two cases (3.4), (3.5) provide the conditions for regularity when $AF - CD > 0$ (i.e., $W < 1$ in canonical form).

We will adopt here the following notation, first introduced in [47]: for any rational function $\rho(A, B, C, D, E, F)$, the corresponding function for the canonical form is denoted by $\rho_c(U, V, W)$, where the expressions are obtained by replacing $A, B, C, \ldots$ by $1, 2U\sqrt{W}, W, \ldots$, except for a multiplicative positive scaling. (See Table 6.10 for a list of the commonly used functions.) Also, for any rational function $\rho_c(U, V, W)$, we define $\rho_c^*(U, V, W) = \rho_c(U, V, W^{-1})$ and $\rho_c^\dagger(U, V, W) = \rho(V, U, W)$. It is finally observed in [47] that $\sigma_c^* = \sigma_c^\dagger = \sigma_c$ and $K_c^* = K_c^\dagger = K_c$.

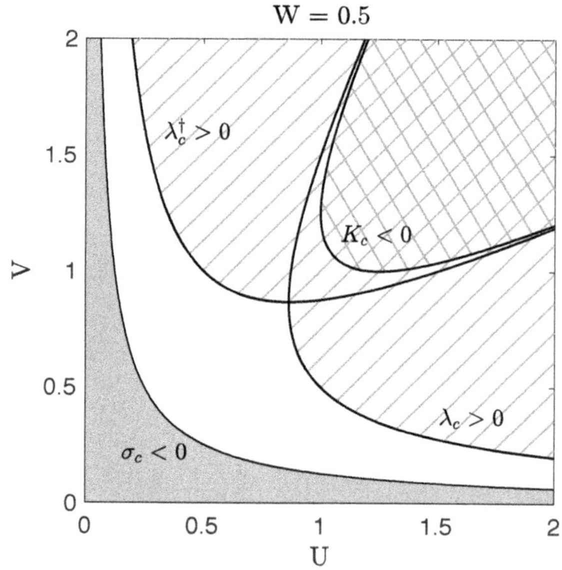

Figure 3.1. Regular region *for* $W = 0.5$. *The expressions for* K_c, σ_c, λ_c, *and* $\lambda_c^\dagger$ *can be found in Table* 6.10. *The shaded region* ($\sigma_c < 0$) *corresponds to nonpositive-real impedances, while the hatched region* ($\lambda_c \geq 0$ *or* $\lambda_c^\dagger \geq 0$) *is the regular region.*

3.2 ▪ Reichert's theorem

Reichert's theorem [67, 48] establishes that the class of impedances which can be realized using two reactive elements is not increased by using more than three resistors. The theorem can be stated as follows.

Theorem 3.2. *Any biquadratic which can be realized using two reactive elements and an arbitrary number of resistors can be realized with two reactive elements and three resistors.*

An immediate consequence of this is that any impedance which can be realized with two reactive elements and an arbitrary number of resistors can also be realized by a network in the Ladenheim catalogue. Since the Ladenheim catalogue does not cover all the possible positive-real biquadratic impedances (realizability regions for all the networks in the catalogue will be illustrated in Chapter 6), a consequence of Reichert's theorem is that some positive-real functions will necessarily have an RLC realization with more reactive elements than the degree of the impedance.

This result was first proven by Reichert in a German language publication [67], using a complicated topological argument. The proof was later reworked in [48] and new lemmas were provided to expand and clarify the main topological argument. More recently, an alternative proof based on a result in [16] was provided in [105]. We provide here an outline of the proof given in [48].

We first note the necessary and sufficient conditions of Section 3.1.1 for a biquadratic to be regular, and the corresponding realizability region plotted in Figure 3.1 for $W = 0.5$. It was shown in [47] that any regular biquadratic can be realized by one of six SP networks with two reactive elements and three resistors. It was also shown that, among all the networks with two reactive elements and three resistors, only two realize nonregular biquadratics. The overall realizability region of the class is shown in Figure 3.2 for $W = 0.6$. From the figure it is clear that the nonregular biquadratics corresponding to the region Γ are not realizable by a two-reactive, three-resistor network. The proof in [48] aims to show that the region Γ is not realizable even if an arbitrary number of resistors is allowed.

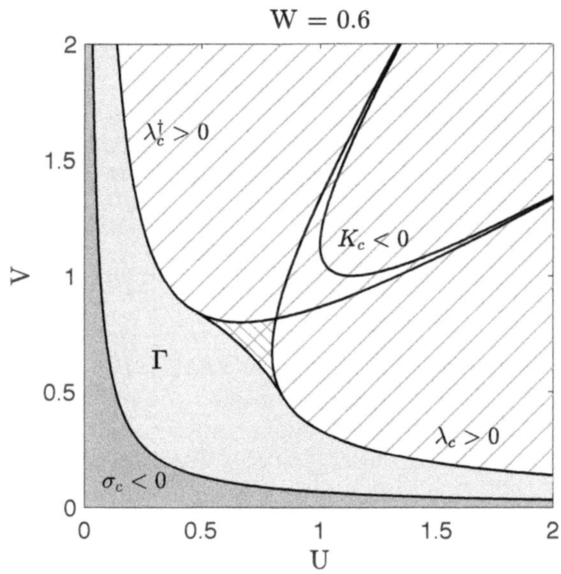

Figure 3.2. *Overall realizability region (hatched) for networks in the Ladenheim catalogue for $W = 0.6$. We note that the realizability region includes some nonregular biquadratics (crossed region). The dark grey region corresponds to nonpositive-real biquadratics, while the light grey region Γ corresponds to nonregular positive-real biquadratics.*

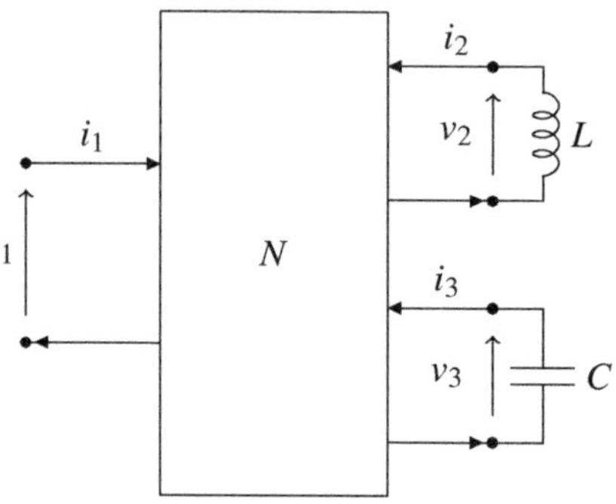

Figure 3.3. *Network N with one inductor, one capacitor, and an arbitrary number of resistors.*

Given a network with one inductor, one capacitor, and an arbitrary number of resistors, it can always be arranged in the form of Figure 3.3, following the reactance extraction method of Section 2.7. Networks with two inductors or two capacitors will always lead to a regular biquadratic (as already mentioned in Section 3.1.1) and are therefore not of interest for the proof.

For the network in Figure 3.3 we can write under mild conditions

$$\begin{pmatrix} v_1 \\ v_2 \\ v_3 \end{pmatrix} = \begin{pmatrix} X_1 & X_4 & X_5 \\ X_4 & X_2 & X_6 \\ X_5 & X_6 & X_3 \end{pmatrix} \begin{pmatrix} i_1 \\ i_2 \\ i_3 \end{pmatrix} =: X \begin{pmatrix} i_1 \\ i_2 \\ i_3 \end{pmatrix}, \tag{3.13}$$

where X is a positive semidefinite, real matrix. In order for X to be realizable as a purely resistive network it is necessary that it is *paramount* (see Section 2.7). Eliminating i_2, i_3, v_2, v_3 from (3.13), and knowing that the driving-point impedance of the network is the ratio of the Laplace transforms of v_1 and i_1, we obtain

$$Z(s) = \frac{(X_1X_3 - X_5^2)s^2 + \left(\frac{X_1}{C} + \frac{\det(X)}{L}\right)s + \frac{X_1X_4 - X_4^2}{LC}}{X_3s^2 + \left(\frac{1}{C} + \frac{X_2X_3 - X_6^2}{L}\right)s + \frac{X_2}{LC}}. \tag{3.14}$$

Using an equivalent characterization of Auth [2] for the impedance (3.14), and by equating this parametrization to the biquadratic canonical form (3.12), the necessary condition on X is translated into a set of necessary conditions involving U, V, and W, which can be interpreted in a topological way. The main part of the proof is based on this topological interpretation and is supported by a series of lemmas. In particular, it is shown in [48] that if it is postulated that there exists a region inside Γ which satisfies the necessary conditions for X to be realizable, then this leads to a contradiction. Therefore, the nonregular region Γ in Figure 3.2 is not realizable even if an arbitrary number of resistors are added to the network.

3.3 ▪ Algebraic criteria for circuit realizations

In [28] Foster stated the following fact for biquadratic impedances without proof: the sign of the resultant K of the numerator and denominator determines whether the reactive elements in a minimally reactive realization are of the same type or of opposite type. This fact was highlighted in a more formal statement by Kalman [51], who suggested that a general proof was urgently needed. A proof was later provided in [41] together with a generalization to impedances of any order.

The approach of [41] made use of Youla and Tissi's reactance extraction approach [102] together with classical results from matrix theory. A series of equivalent criteria are presented in [41] and are expressed in terms of the rank, signature, or number of permanences/variations in the sign of the determinants of certain matrices. In particular, conditions are given on the Hankel matrix (whose entries are defined from the Laurent expansion of $Z(s)$), on an extended Cauchy index, and on the Sylvester and Bezoutian matrices (whose entries can be obtained from the coefficients of the numerator and denominator of $Z(s)$). We present here the conditions given in terms of the Sylvester matrices.

Given a positive-real impedance function

$$Z(s) = \frac{a_n s^n + a_{n-1}s^{n-1} + \cdots + a_0}{b_n s^n + b_{n-1}s^{n-1} + \cdots + b_0}, \tag{3.15}$$

let the associated *Sylvester matrices* be defined as

$$S_{2k} = \begin{pmatrix} b_n & b_{n-1} & \cdots & b_{n-k+1} & b_{n-k} & \cdots & b_{n-2k+1} \\ a_n & a_{n-1} & \cdots & a_{n-k+1} & a_{n-k} & \cdots & a_{n-2k+1} \\ 0 & b_n & \cdots & b_{n-k+2} & b_{n-k+1} & \cdots & b_{n-2k+2} \\ 0 & a_n & \cdots & a_{n-k+2} & a_{n-k+1} & \cdots & a_{n-2k+2} \\ \vdots & \vdots & \ddots & \vdots & \vdots & \ddots & \vdots \\ 0 & 0 & \cdots & b_n & b_{n-1} & \cdots & b_{n-k} \\ 0 & 0 & \cdots & a_n & a_{n-1} & \cdots & a_{n-k} \end{pmatrix}$$

for $k = 1, \ldots, n$, where $a_i = 0$ and $b_i = 0$ if $i < 0$. Then the following theorem holds.

Theorem 3.3. *Let $Z(s)$ in (3.15) be the impedance of an RLC network containing exactly p inductors and q capacitors (with $p + q = n$). Then $\det(S_{2n}) \neq 0$ and*

$$q = \textit{permanences of sign in the sequence } \big(1, \det(S_2), \det(S_4), \ldots, \det(S_{2n})\big),$$
$$p = \textit{variations of sign in the sequence } \big(1, \det(S_2), \det(S_4), \ldots, \det(S_{2n})\big).$$

Signs for any subsequence of zeros $\det\big(S_{2(k+1)}\big) = \det(S_{2(k+2)}) = \cdots = 0$, where $\det(S_{2k}) \neq 0$, are assigned as follows:

$$\operatorname{sign}\big(\det(S_{2(k+j)})\big) = (-1)^{\frac{j(j-1)}{2}} \operatorname{sign}\big(\det(S_{2k})\big).$$

We now apply Theorem 3.3 to a biquadratic impedance in the notation of (3.1). The associated Sylvester matrices are

$$S_4 = \begin{pmatrix} D & E & F & 0 \\ A & B & C & 0 \\ 0 & D & E & F \\ 0 & A & B & C \end{pmatrix}, \qquad S_2 = \begin{pmatrix} D & E \\ A & B \end{pmatrix},$$

and it is easily verified that the resultant $K = -\det(S_4)$ has the expression given in (3.3), while $\det(S_2) = BD - AE$. The number of inductors and capacitors in the network is therefore determined by the number of permanences and variations of sign in the sequence $(1, BD - AE, -K)$. When $K > 0$ there will therefore always be one inductor and one capacitor regardless of the sign of $BD - AE$, while when $K < 0$ one can differentiate between two different cases, as summarized in Table 3.1. We note that in Table 3.1 the sign of $-(AF - CD)$ is equivalently used instead of sign$(BD - AE)$, as in [28].

Table 3.1. *Number of reactive elements (# inductors, # capacitors) in a minimally reactive realization of a biquadratic impedance. The case $K < 0$, $AF - CD = 0$ cannot occur, since the two conditions would imply $K = 0$.*

	$AF - CD > 0$	$AF - CD < 0$	$AF - CD = 0$
$K > 0$	(1,1)	(1,1)	(1,1)
$K < 0$	(2,0)	(0,2)	–

3.4 ▪ The behavioral approach to passivity

The long-standing question of the apparent nonminimality of the Bott–Duffin networks, which was mentioned in Section 2.5, has recently prompted a fresh treatment of the driving-point behavior of RLC networks [43] and, more generally, a new analysis of passive behaviors [38], using Willems' behavioral approach [66, 99].

Given an RLC n-port network, let $\mathbf{i}$ and $\mathbf{v}$ denote the vectors of length n of driving-point currents and voltages. Then the driving-point behavior $\mathcal{B}$ of the network can be described as a linear time-invariant differential behavior, i.e., the set of solutions to a system of differential equations of the form

$$P_0\mathbf{i} + P_1\frac{d}{dt}\mathbf{i} + \cdots + P_m\frac{d^m}{dt^m}\mathbf{i} = Q_0\mathbf{v} + Q_1\frac{d}{dt}\mathbf{v} + \cdots + Q_m\frac{d^m}{dt^m}\mathbf{v}, \tag{3.16}$$

where $m \geq 0$, $P_0, \ldots, P_m$ and $Q_0, \ldots, Q_m$ are square real matrices of dimension n, and $\mathbf{i}$ and $\mathbf{v}$ are assumed to be locally integrable functions [37]. The system (3.16) can be written more compactly as

$$P\left(\frac{d}{dt}\right)\mathbf{i} = Q\left(\frac{d}{dt}\right)\mathbf{v}, \tag{3.17}$$

where $P(\xi) = P_0 + P_1\xi + \cdots + P_m\xi^m$, and similarly for $Q(\xi)$. The following definition of a *passive* system in terms of its behavior was introduced in [100] and later adapted in [38].

Definition 3.4. *The system described by the behavior $\mathcal{B}$ in* (3.17) *is passive if for any given $(\mathbf{i}, \mathbf{v}) \in \mathcal{B}$ and $t_0 \in \mathbb{R}$ there exists a $K \in \mathbb{R}$ (dependent on $(\mathbf{i}, \mathbf{v})$ and t_0) such that if $(\hat{\mathbf{i}}, \hat{\mathbf{v}}) \in \mathcal{B}$ satisfies $(\hat{\mathbf{i}}(t), \hat{\mathbf{v}}(t)) = (\mathbf{i}(t), \mathbf{v}(t))$ for all $t < t_0$, then*

$$\int_{t_0}^{t_1} \hat{\mathbf{i}}(t)^T \hat{\mathbf{v}}(t)\, dt > -K$$

for all $t_1 \geq t_0$.

In words the definition says that, given an element of the behavior, i.e., a trajectory $(\mathbf{i}, \mathbf{v})$ which satisfies the representation (3.17), there is a bound K on the energy which can be extracted from the network from the present time t_0 onwards. This bound depends on the specific past trajectory up to t_0 but applies to any future trajectory after t_0. We note that this definition is different from the classic notion of passivity of a one-port given in Definition 2.7 but still formalizes the underlying property that it is not possible to extract unlimited energy from the network.

In the case of one-port RLC networks, the behavior of the system takes the form

$$p\left(\frac{d}{dt}\right) i = q\left(\frac{d}{dt}\right) v, \tag{3.18}$$

where $p(s)$ and $q(s)$ are polynomials in s with real coefficients. The driving-point impedance of the network is given by $Z(s) = p/q$ and we would expect the condition of positive-realness of Z to be equivalent to Definition 3.4, which would be in agreement with Sections 2.3 and 2.4. In general, however, for a behavior of the form (3.18) to be passive it is necessary but not sufficient that the function p/q is positive-real: this is due to the possibility of common roots between p and q, which arise when the behavior is not controllable [37]. The following theorem provides a necessary *and sufficient* condition for the behavior of a multiport network to be passive [38].

Theorem 3.5. *Let $\bar{\lambda}$ denote the complex conjugate of λ and $\mathbb{R}^n[s]$ the vectors of dimension n whose entries are polynomials in s with real coefficients. The system* (3.17) *is passive if and only if the following three conditions hold:*

1. $P(\lambda)Q(\bar{\lambda})^T + Q(\lambda)P(\bar{\lambda})^T \geq 0$ *for all λ in the closed right half-plane;*
2. $\operatorname{rank}([P \;\; -Q](\lambda)) = n$ *for all λ in the closed right half-plane;*
3. *if $\mathbf{r} \in \mathbb{R}^n[s]$ and $\lambda \in \mathbb{C}$ satisfy $\mathbf{r}(s)^T\left(P(s)Q(-s)^T + Q(s)P(-s)^T\right) = 0$ and $\mathbf{r}(\lambda)^T[P \;\; -Q](\lambda) = 0$, then $\mathbf{r}(\lambda) = 0$.*

In the case of a one-port characterized by the behavior (3.18) with $Z(s) = p/q$ we have $n = 1$, and the first condition in Theorem 3.5 corresponds to positive-realness of Z, the second condition establishes that there are no pole-zero cancellations in the function p/q in the closed right half-plane, and the third condition implies that, in the lossless case (where $Z(s)+Z(-s) \equiv 0$), pole-zero cancellations are not allowed even in the left half-plane, meaning that p and q must be coprime [36]. It is therefore not sufficient that the function p/q is positive-real (condition 1) for the behavior to be passive, since there might be pole-zero cancellations in p/q, which do not satisfy the other two conditions of the theorem.

Chapter 4

Mechanical networks

Many of the modern developments in the field of network synthesis were motivated by the introduction of a new fundamental component for mechanical control, the inerter [78], alongside the spring and the damper. This new network element provided a way to realize passively any positive-real mechanical admittance or impedance and therefore to directly exploit the wealth of results from electrical network synthesis. Since its introduction in the early 2000s, the inerter has been successfully employed in passive suspensions in motorsport including Formula One cars [15] and is being extensively researched for a wide range of other applications, some of which will be discussed in Section 4.2.

We consider in this chapter *mechanical networks* consisting of a finite interconnection of mechanical elements. Analogous to the case of electrical networks, a *port* in a mechanical system is a pair of terminals to which an equal and opposite force F is applied with a relative velocity v between the terminals. The sign convention is shown in Figure 4.1.

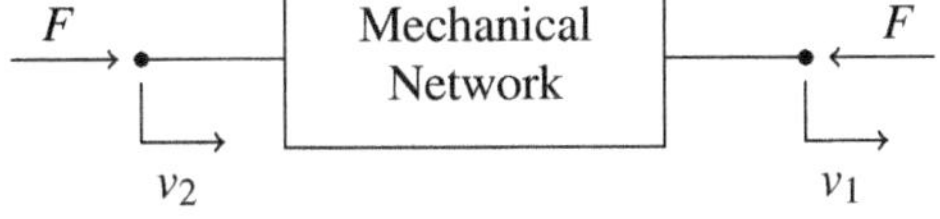

Figure 4.1. *One-port (two-terminal) mechanical element or network, with the convention that a positive F gives a compressive force and a positive $v = v_2 - v_1$ corresponds to the terminals moving towards each other.*

4.1 ▪ Network analogies and the inerter

There are two standard analogies between electrical and mechanical systems. Historically the first of these is the so-called *force-voltage* analogy (in which force is analogous to voltage and velocity is analogous to current) as evidenced in the terminology electromotive force. The *force-current* (also known as *mobility*) analogy was subsequently introduced by Firestone [24] (see also [20, 34]), who also introduced the concepts of through and across variables. A *through variable* has the same value at the two terminals of the element (e.g., force and current), while an *across variable* is given by a difference of the value at the terminals (e.g., velocity, voltage). Insight on whether a variable is through or across can be gained by considering how measurements of such a variable are taken: through variables require a single measurement point (and typically require the system to be severed at that point), while across variables are measured as a difference between two measurement points (without having to break into the system). This framework allowed analogies to be extended to any dynamical system where through and across measurements can be obtained, such as thermal, fluid, and acoustic systems [55, 75]. We mention that there is a corresponding analogy between electrical networks and mechanical systems in rotational form.

In the force-current analogy between mechanical and electrical networks, force (respectively, velocity) corresponds to current (respectively, voltage) and a fixed point in an inertial frame of reference corresponds to the electrical ground [75]. In this analogy the element correspondences are often stated in the following form:

$$\begin{aligned} \text{spring} &\quad \longleftrightarrow \quad \text{inductor} \\ \text{damper} &\quad \longleftrightarrow \quad \text{resistor} \\ \text{mass} &\quad \longleftrightarrow \quad \text{capacitor} \end{aligned}$$

The correspondence is perfect in the case of the spring and damper, but there is a restriction in the case of the mass due to the fact that it has only one independently movable terminal, the center of mass. Since the force-velocity relationship relates the acceleration of the center of mass to a fixed point in the inertial frame, the mass element is, in effect, analogous to a grounded capacitor. This means that, using the classical analogy described above, an RLC circuit may not have a direct mass-spring-damper mechanical analogue, given that in the electrical domain capacitors are not in general required to be grounded.

To complete the analogy, a new two-terminal device, the "inerter," was introduced in [78], with the property that the applied force at its terminals is proportional to the *relative* acceleration between them. The constant of proportionality is called *inertance* and has the units of kilograms. A table of the circuit symbols of the six basic mechanical and electrical elements, with the inerter replacing the mass, is shown in Figure 4.2, along with their defining equations. In order to justify the introduction of the inerter as an ideal modeling element it should be possible to physically realize inerters which satisfy a number of practical requirements: it should be a two-terminal device which allows sufficient linear travel, which does not need to be attached to any fixed point, which works in any spatial orientation and motion, and which has a mass that is small compared to the elements to which it is connected and independent of the desired value of inertance. Many different physical embodiments of the inerter which satisfy these conditions to a sufficient degree of approximation were devised, ranging from mechanical devices like the rack and pinion inerter and the ballscrew inerter to hydraulic mechanisms using a gear pump [78, 79, 64, 92] and the fluid inerter [86].

An embodiment of the inerter in rotational form was also given in [79], thus completing (along with the rotary spring and damper) the mechanical-electrical analogy in rotational form. In this case the two terminals of the device can independently rotate about a common axis and an equal and opposite torque is applied at the terminals. The relation between the torque at the terminals and their relative angular displacement, velocity, and acceleration gives the defining equations of the rotary spring, damper, and inerter, respectively.

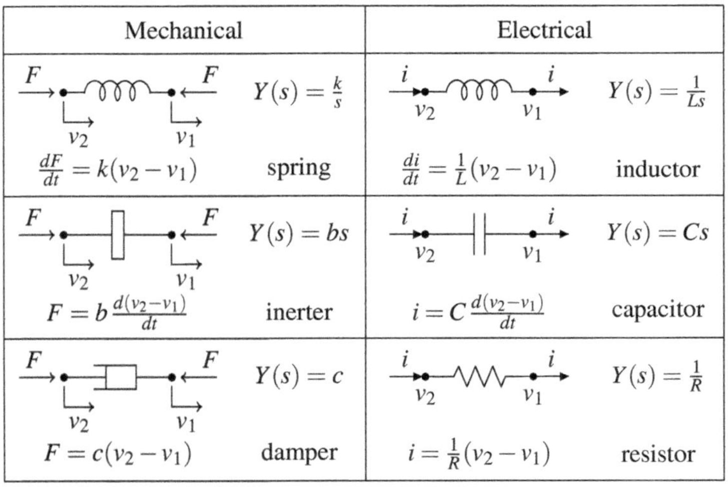

Figure 4.2. *The basic mechanical and electrical circuit elements, with their symbols, admittance function, and defining equations.*

In the force-current analogy, the *mechanical impedance* is taken to be the ratio between the Laplace transforms of velocity and force, i.e., between an across variable and a through variable (with the *admittance* being the reciprocal of the impedance). We also note that the force-current analogy respects the manner of interconnection; therefore in order to obtain the electrical or mechanical equivalent of a network it is sufficient to replace each element with the corresponding element in the other domain, while maintaining the same network topology [24].

The most significant consequence of the introduction of the inerter is the possibility to exploit the full freedom of passive network synthesis to synthesize mechanical impedances. The Bott–Duffin theorem (see Theorem 2.9) established that any positive-real function can be realized by an electrical network containing resistors, inductors, and capacitors only; with the introduction of the inerter it is possible to find the mechanical equivalent of any given electrical network, and therefore the following result can be stated.

Theorem 4.1. *Any positive-real rational function can be realized as the impedance of a one-port mechanical network containing springs, dampers, and inerters only.*

4.2 ▪ Applications of mechanical network synthesis

With the introduction of the inerter, problems of passive mechanical control can be split into two subproblems: (i) the design of a suitable positive-real mechanical admittance function $Q(s)$ (i.e., a control systems design problem, for example, the optimization of a given performance index) and (ii) the synthesis of a physical mechanical network with admittance $Q(s)$ (for which the passive network synthesis methods described in Chapter 2 and in the remainder of this monograph can be employed, as explained in Section 6.5.1). Since Theorem 4.1 guarantees that any function can be realized as the impedance of a physical mechanism as long as it is positive-real, this design paradigm offers much more power and flexibility than traditional design methods.

The potential in being able to fully exploit passive network synthesis methods in the field of mechanical control will be highlighted in Section 4.2.2 for the vehicle suspension problem. In addition to vehicle suspension design [65, 73, 78], other applications of the inerter and of passive network synthesis include control of motorcycle steering oscillations [21, 22, 50], rail suspensions [44, 45, 95], building suspensions [91, 101], and vibration suppression for machine tools [94] or for support and isolation of structures [76]. The common feature of these and other novel applications is the relatively low complexity of the passive networks being considered—which is a motivation for the present work.

4.2.1 ▪ Vibration absorption

The vibration absorption problem is a classical problem whose objective is to minimize the vibrations of a mass M connected to a structure that is subject to oscillations at a constant frequency ω_0, as illustrated in Figure 4.3. Traditionally this problem is solved with a tuned spring-mass system mounted on the mass itself—which is known as a dynamic vibration absorber (DVA). It was shown in [78] that inerters can be used in an alternative design that does not require any element to be mounted on the mass.

The traditional design is shown in Figure 4.4(a) and consists of a spring-mass system mounted on the mass M, as well as a parallel spring-damper connecting the mass M to the vibrating structure. It can be easily shown that, when $F_L = 0$, the displacement of the mass M in response to a sinusoidal disturbance z asymptotically goes to zero. On the other hand, the oscillations of the additional mass m can be quite large at steady state, and a large mass m may be needed to avoid this. With the introduction of the inerter any positive-real mechanical admittance $Q(s)$ in Figure 4.3 may be chosen and then physically realized, and it is shown in [78] that similar performance may be achieved in suppressing the oscillations of the mass M by using the spring-damper-inerter mechanism shown in Figure 4.4(b), with the advantage that no additional mass is mounted on the mass M.

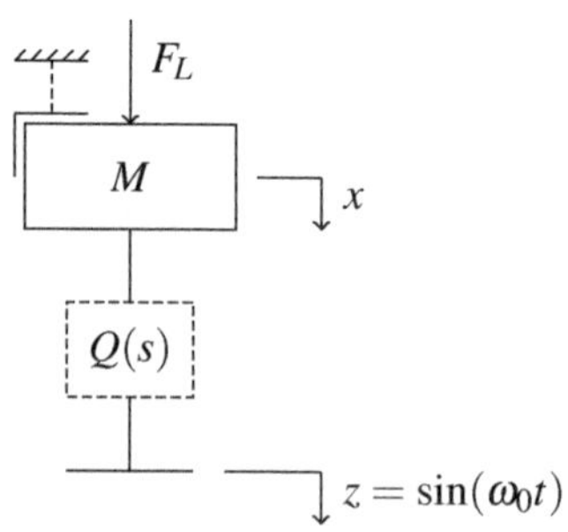

Figure 4.3. *Problem setup for the vibration absorption problem. The mass M may be subject to a force F_L. The displacements of the mass and of the vibrating structure are denoted as x and z, respectively.*

Recent applications of the inerter in vibration suppression include milling machine tools [94] and an isolator for Michelangelo's Pietà Rondanini statue [76]. In the first case, an inerter is included in passive mechanical networks which aim at reducing cutting vibration, as an alternative to active techniques and traditional tuned mass-damper (TMD) systems [29] (namely a spring-mass-damper oscillator mounted on the mass

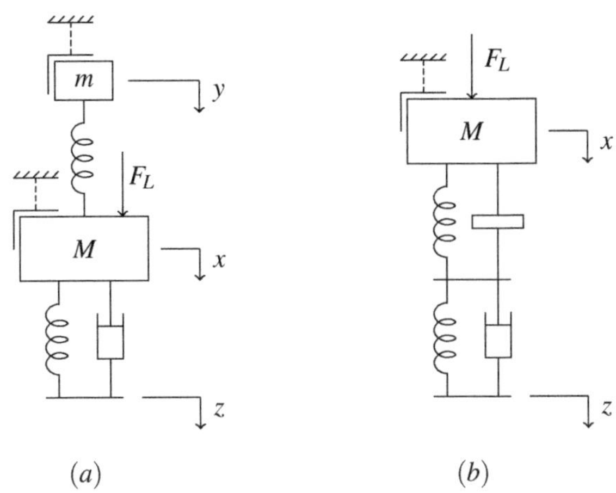

Figure 4.4. (a) *Traditional design for a DVA;* (b) *new design for the vibration absorption problem which includes an inerter.*

whose vibrations are to be controlled, much like in Figure 4.4(a)). The possibility of including an inerter in two distinct suspension layouts is explored, which is shown to improve the manufacturing performance according to a series of performance indices. A similar approach is adopted in [76], where a passive vibration isolator which includes inerters is devised, to protect the statue against earthquake vibrations and traffic vibrations in central Milan. It is shown there that the tuned mass-damper-inerter (TMDI) solution results in a significant reduction in vibrations (both in the horizontal and the vertical directions) compared to a traditional TMD system.

We also note that TMDI systems have seen wide application in the case of vibration absorption in buildings and other engineering structures, as will be discussed in more detail in Section 4.2.5.

4.2.2 ▪ Vehicle suspensions

Another important application of mechanical network synthesis, which was the initial motivating application for the inerter, is in vehicle suspensions. (See [15, 80] for an account of the use of the inerter in motorsport.) In general, the objective in suspension design is to provide a comfortable ride and good handling, considering that two main disturbance categories may act on the vehicle, road disturbances and driver inputs, the latter being sometimes modeled as load disturbances. A simple quarter-car vehicle model like the one shown in Figure 4.5(a) can be used to formulate the problem (although full car models have also been considered, e.g., [85]). It consists of a sprung mass m_s, an unsprung mass m_u, and a tire with spring stiffness k_t. The suspension strut is assumed to be a passive mechanical network with admittance $Q(s)$ which provides an equal and opposite force on the masses. The electrical analogue of the quarter-car model is shown in Figure 4.5(b).

A fully active suspension offers advantages in terms of design freedom (e.g., decoupling of ride and handling behavior [83, 84]), but at the price of increased complexity and expense. On the other hand, a traditional passive suspension strut consisting of a spring in parallel with a damper is easily realized but is quite limited in terms of the available mechanical admittances (namely, the poles and zeros alternate on the negative

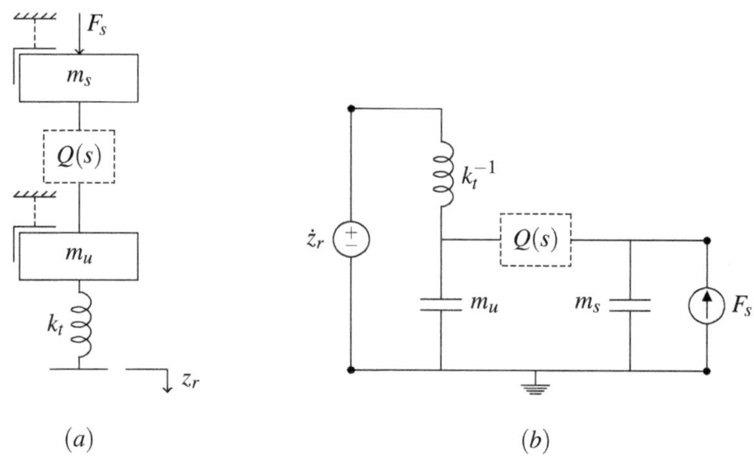

Figure 4.5. (a) *Quarter-car vehicle suspension model with a load F_s on the sprung mass and vertical velocity $\dot{z}_r$ of the road-tire contact as exogenous inputs;* (b) *electrical analogue of the quarter-car model, obtained by using the force-current analogy: the two masses become grounded capacitors, while the exogenous inputs (F_s and $\dot{z}_r$) become a current and a voltage source, respectively.*

real axis, thus resulting in a frequency response phase characteristic that is always lagging). Considering instead any given positive-real admittance $Q(s)$ offers much greater possibilities than a conventional passive strut, while maintaining the advantages of having a suspension system without internal power source.

In the case of passive devices, the suspension design problem can be formulated as an optimal control problem over certain classes of positive-real admittances $Q(s)$ [65, 85] (typically low-order positive-real admittances), as illustrated in Figure 4.6 from a systems control perspective. Typical performance criteria in the optimization problem include ride comfort, road holding, handling, and the ability to withstand external loads on the sprung mass.

In [73], six suspension struts of fixed structure are selected, and global optima are obtained analytically for ride comfort and tire grip performance measures as a function of the quarter-car model parameters. This approach builds on the results presented in [85] and has the additional advantage that the numerical optimization does not need to be repeated if different vehicle parameters are considered.

A different optimization problem is considered in [82], namely the problem to maximize the least damping ratio among the poles of the system, over various classes of low-order admittances. A number of simple suspension networks with fixed structure (some including inerters) are considered for this purpose. Optimal solutions are found for the least damping ratio as a function of key parameter ratios and the reduction in oscillatory behavior is compared for the networks considered.

Finally, a new design paradigm which aims at reducing the limitations of considering suspension networks with a fixed structure is described in [104]—the so-called structure-immittance approach. This approach seeks an optimal network configuration and element values among all networks with a certain number of springs, dampers, and

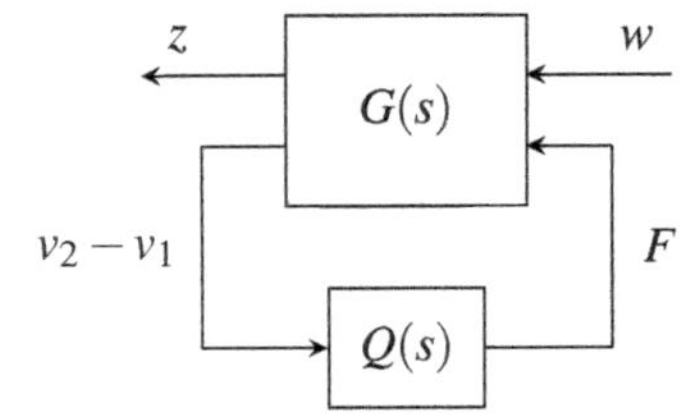

Figure 4.6. *Control systems paradigm for the suspension design problem. Positive-realness of the suspension device $Q(s)$ can be characterized through the positive-real lemma (with a state-space approach). The input w to the generalized plant $G(s)$ represents exogenous disturbances, while z represents the outputs to be controlled, like tire force, sprung mass acceleration, etc.*

inerters, while also allowing constraints on the element values. The case of networks with one damper, one inerter, and at most two springs is analyzed in [104].

4.2.3 ▪ Control of motorcycle steering oscillations

An application of the inerter in rotational form has been considered in a mechanical steering compensator for high-performance motorcycles [21, 22, 50]. Studies have shown that there are two important modes in motorcycle dynamics which, under certain conditions, may become very underdamped or even unstable: the *wobble* and the *weave* modes [58]. The former is a steering oscillation, while the latter is a fishtailing-type motion which involves roll and yaw. Traditionally, the wobble is stabilized at high speed by means of a steering damper, which, however, has a destabilizing effect on the weave mode.

It is shown in [21, 22] that an inerter (in its rotational form) can be employed in the steering compensator to simultaneously control the two modes and improve overall performance. The problem is described from a control systems point of view in Figure 4.7, similarly to Figure 4.6. We note that in the conventional control system, the compensator $K(s)$ consists of a steering damper (i.e., a constant of proportionality between steering torque and steering velocity). More general mechanical admittances are considered in [21, 22] as steering compensators, in an effort to obtain a mechanism that behaves like a damper over the wobble frequency range and like an inerter over the weave mode frequencies. A number of simple spring-damper-inerter networks of up to four components are taken as candidate networks and an optimization is performed over the network parameters, in order to identify the most beneficial layout and the optimal element values. A further candidate admittance given by a general biquadratic positive-real function is considered, and the optimization is carried out over the impedance coefficients in this case. The optimization is performed according to two criteria, one in a time-domain framework and the other in a frequency-domain framework:

1. maximizing the motorcycle's smallest damping ratio, among all modes (similarly to the approach [82] discussed in Section 4.2.2);
2. introducing an $\mathcal{H}_\infty$ response measure which aims at minimizing the worst-case gain from road-displacement disturbances to steering angle, together with a penalty on the closeness of the open-loop Nyquist plot to the -1 point.

It turns out that the use of the inerter significantly improves performance according to both measures compared to conventional steering dampers. It also turns out that the optimal configuration, among the networks of the structures considered, is a series spring-damper-inerter network. A physical mechanical steering compensator in rotary form was prototyped at the Cambridge University Engineering Department for this purpose [63, 50].

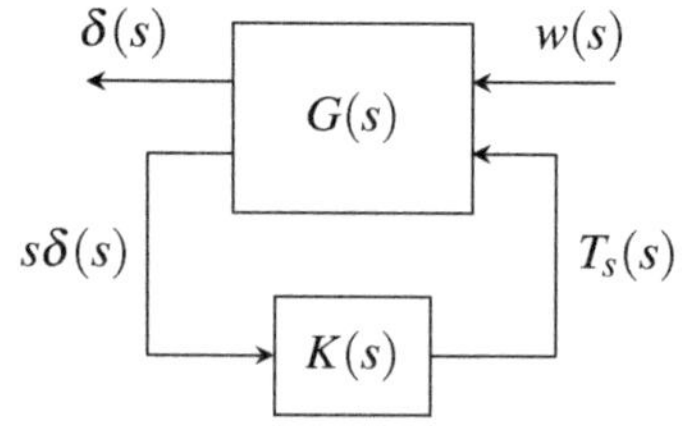

Figure 4.7. *Block diagram describing the motorcycle steering control problem. The plant $G(s)$ is given by the linearized motorcycle model, $K(s)$ is the steering compensator, and the input $w(s)$ represents vertical road-displacement forcing, while $T_s(s)$ is the steering torque. The signals $\delta(s)$ and $s\delta(s)$ are the Laplace transforms of the steering angle and angular velocity, respectively.*

4.2.4 ▪ Train suspensions

A more recent application of the inerter is in the suspension systems of railway vehicles. Studies have been made, starting from [96], to show how the inerter can improve performance in different ways, ranging from ride quality to the maximum forward speed attainable [44, 45, 95, 46, 14]. Train models similar to the one shown in Figure 4.8, or variations thereof, are usually considered and are based on the models of Goodall and coworkers (see, for example, [31, 9]). Only half of the car and one bogie which carries two wheelsets is modeled in Figure 4.8, though full-train models with two bogies can also be employed [93] as well as multicar models (see [14] for a study on the effects of having a number of connected cars on system stability and performance). Typically, railway wheelsets comprise two coned wheels joined by a solid axle, which allow natural centring on the track. Different numbers of degrees of freedom (DOF) may result in the model, depending on which modes are neglected (e.g., vertical or roll). The suspensions between the wheelsets and the bogie are termed *primary* suspensions, while the suspensions between the bogie and the vehicle body are termed *secondary*. In the plan-view model of Figure 4.8, the longitudinal and lateral suspension systems are given by general mechanical admittances; primary and secondary vertical suspensions (in the z-axis direction) are also usually present.

Traditional rail suspension systems usually consist of a spring-damper parallel connection. Mechanical networks which include inerters can be employed in some or all of the primary and secondary suspension locations to improve performance. In [44, 45, 95], simple networks of fixed structure which include one or two inerters are considered, an optimization is performed over the suspension parameters among the different layouts, and the results are compared with the case of a conventional spring-damper suspension. The objective of the optimization may vary in the different cases.

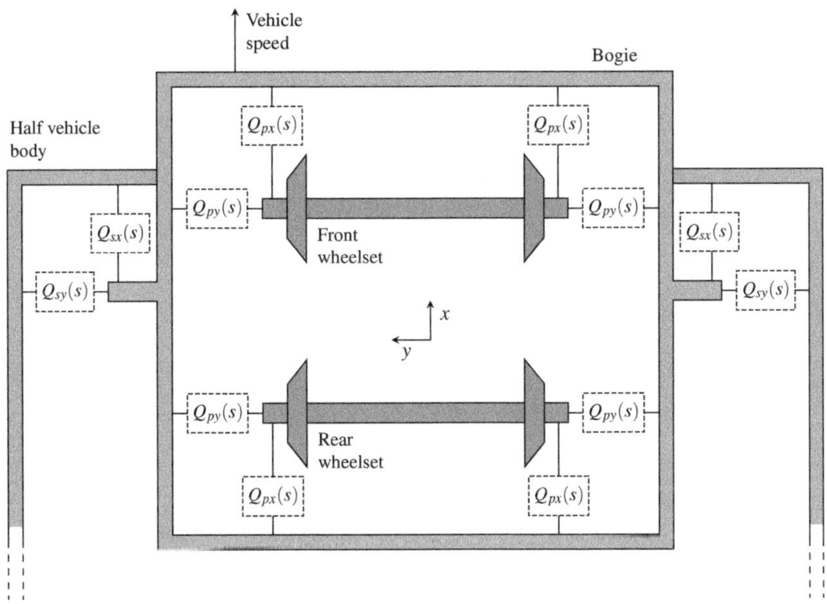

Figure 4.8. *Plan-view model of a train consisting of a half vehicle body, one bogie frame, and two solid axle wheelsets. The primary and secondary vertical suspensions $Q_{pz}(s)$ and $Q_{sz}(s)$ are not shown.*

In [44] a seven-DOF train model is considered, and two simple networks containing one spring, one damper, and one inerter are used as candidates for the secondary lateral suspensions, with parallel spring-damper suspensions employed elsewhere. An optimization is then performed over the suspension parameters with the objective of improving lateral and vertical ride comfort, as well as lateral body movement when curving. In [95] a 16-DOF model is considered and the possibility of using inerters in all primary and secondary suspensions is explored, with the objective of improving the critical speed of the train, i.e., the maximum stable speed. The same basic suspension layouts as in [44] are considered there, and it is concluded that the optimal solution is having a parallel spring-damper-inerter network in all suspension locations. In addition to simple networks with a fixed structure, more general positive-real mechanical admittances of fixed degree are considered in [96], and an optimization is carried out using linear matrix inequality approaches, with two performance indices which take into account passenger comfort and dynamic wheel load. General positive-real admittances of first, second, and third orders are considered in the optimization problem and are then realized using the Bott–Duffin method (see Section 2.5).

Finally, a single-stage vehicle is considered in [45, 46], namely a vehicle without bogies or secondary suspensions, allowing a simpler construction and a smaller weight. The objective in [45] is to improve ride quality through the use of inerters in the lateral and vertical suspensions, given that unsatisfactory ride quality is the main drawback of single-stage vehicles. A six-DOF model is employed, and four candidate suspension systems with one damper and up to two inerters and two springs are considered. The optimization over the suspension parameters over the four layouts shows that ride comfort can be significantly improved at most speeds with the use of inerters. The objective in [46], instead, is to improve curving performance by incorporating the inerter in the lateral suspensions of a single-stage vehicle. In conventional railway vehicles additional yaw stiffness is added in the suspensions in order to reduce oscillations in the horizontal plane caused by the profile of the wheels, at the price of an increased wear of the wheels and rails. It is shown in [46] that the inerter can be used to avoid the additional stiffness with the effect of improving the curving performance without worsening ride comfort and stability.

4.2.5 ▪ Building suspensions

We conclude this chapter by looking at how mechanical networks containing inerters can improve vibration control in building suspensions.

This problem was approached in a similar way to the rail suspension problem in [90, 91]. Buildings with isolators at their base are considered there, with the objective of suppressing vibrations due to earthquakes and traffic (which occur in specific frequency ranges). The model for the one-DOF case shown in Figure 4.9(a) is adopted, which is similar to the problem setup of Figure 4.3, where $Q(s)$ is the positive-real mechanical admittance of one of four candidate layouts containing springs, dampers, and inerters. $\mathcal{H}_2$ and $\mathcal{H}_\infty$ performance measures are defined and the optimization is performed over the suspension parameters. It is then shown that performance can be further increased by adding an additional mass layer to the suspension. This study is then extended to a two-DOF model of the building, where vertical and lateral suspensions of the base are considered.

The multistory case was studied in [57], where the *tuned inerter-damper* (TID) is introduced as an alternative to the traditional TMD for suppressing vibrations in buildings. Traditional TMD systems in building structures usually consist of mass-spring-damper systems like the one shown in Figure 4.10(a) and are most effective when attached to the top of the primary structure whose vibrations are to be controlled. Performance of TMD systems is limited by the amount of mass that can be added to the structure itself. TID suspensions like the one shown in Figure 4.10(b) can be placed instead between adjacent stories and allow better performance to be obtained with smaller mass and device size. It was found in [57] that the best performance is achieved when the TID is placed at the lowest levels, in contrast to the TMD, which is a further advantage.

The TMDI shown in Figure 4.10(c) was introduced and studied in [60, 30]. A multistory primary structure is considered in [60], with the base being excited stochastically. A TMDI system is attached to the top of the building, with the objective of suppressing the fundamental mode of vibration, and it is shown that this is achieved in a more effective way than a classical TMD and with smaller overall mass of the suspension network. The analysis in [30] additionally accounts for uncertainty in the structural model, and shows that performance is improved compared to a TMD, particularly when the TMDI is placed at lower floors.

A general question is addressed in [101] for the vibration suppression problem in a multistory building. The problem setup is shown in Figure 4.9(b), where the masses M represent the stories, in between which passive control devices of impedance $Z(s)$ are introduced. It is shown that it is possible to bound the $\mathcal{H}_\infty$-norm of the transfer functions from the movable point x_0 to any given intermass displacement, independently of the number of stories. A graphical method is also described to design the mechanical impedances $Z(s)$ between the stories in such a way that this $\mathcal{H}_\infty$-norm is no greater than a prescribed value.

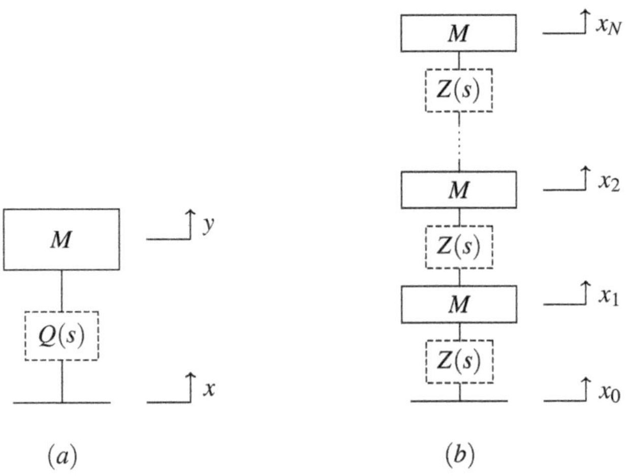

Figure 4.9. (a) *One-DOF rigid-body model of the building, where x and y are the vertical displacements of the base and the building, respectively, and M is the mass of the building;* (b) *vibration suppression in a multistory building, modeled as a chain of N masses of displacements $x_1, \ldots, x_N$ connected by passive mechanical impedances and excited by the movable point x_0.*

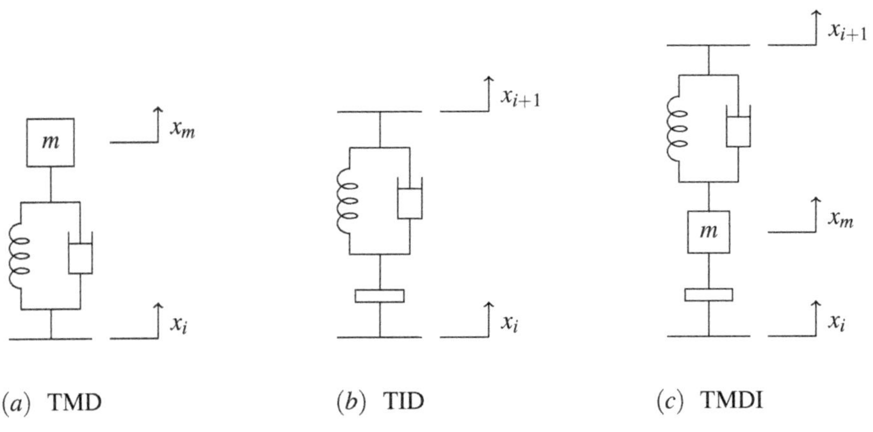

Figure 4.10. (a) *TMD system attached to a movable point x_i;* (b) *TID network connecting two movable points x_i and x_{i+1};* (c) *TMDI system connecting two movable points x_i and x_{i+1}.*

Chapter 5
The enumerative approach to network synthesis

The dissertation of Ladenheim [56] considers all two terminal RLC networks with five elements or less, of which at most two are reactive (inductors or capacitors), and which do not simplify to networks with fewer elements by known network transformations. Considering networks with one reactive element is a trivial case, while the problem with three or more reactive elements is very complex. Hence, the class considered by Ladenheim is the first nontrivial, but still significant, class of networks to analyze. The impedances realized by such networks are biquadratics of the form

$$Z(s) = \frac{As^2 + Bs + C}{Ds^2 + Es + F}, \tag{5.1}$$

where $A, B, C, D, E, F \geq 0$. Ladenheim's derivation and analysis of the canonical set involves the following steps:

- All possible basic graphs with at most five branches are listed and all 148 essentially distinct networks which can be built from these graphs are found. Simple transformations allow some networks to be reduced to equivalent ones with fewer elements. In this way the set is reduced to 108 distinct networks, which may be viewed as the "canonical" set (see Section 5.1 for a detailed derivation of the set).
- Ladenheim then computes the impedance of all 108 networks (starting from the one-element networks, up to the much more interesting five-element networks), i.e., the explicit form of coefficients A, B, C, D, E, F in (5.1) in terms of resistances, capacitances, and inductances.
- An attempt is made on the inverse problem; namely, expressions are stated for the inductances, capacitances, and resistances for each network in terms of the coefficients A, B, C, D, E, F.
- A basic attempt at grouping some of the networks is then performed.

A fundamental question which is left open by Ladenheim's work is that no conditions are given on the coefficients A, B, C, D, E, F which guarantee that the expressions for the inductances, capacitances, and resistances are real and positive. Deriving such necessary and sufficient conditions is one of the major tasks of the present work, which will allow the structure and interrelationships within the catalogue to be illuminated. In preparation for this task, in the next section we will expand and rework the procedure to obtain the canonical set.

5.1 ▪ Definition and derivation of the Ladenheim catalogue

The first step in the derivation of the canonical set is to list all the connected graphs with at most five edges and two special vertices (the external terminals of Figure 2.1). These graphs are enumerated in Appendix B (see also [70, 26, 56]). They are first grouped based on the number of branches and, within each group, based on the number of vertices, and further according to the type of network (as defined in Section 2.1): for graphs A . . . V simple series-parallel (SSP) graphs appear before series-parallel (SP) graphs, which in turn appear before bridge graphs; for graph duals $\mathtt{A}^{\mathrm{d}} \ldots \mathtt{U}^{\mathrm{d}}$ the order is reversed; SSP or SP graphs with the same number of branches and vertices are further ordered, with essentially series graphs appearing before essentially parallel.

The next step is to populate each branch with a resistor or a reactive element to obtain all the *essentially distinct* RLC networks. Networks that are not essentially distinct are related by the operations of deformation, separation, and series interchange (see [26]). The concept is formalized in graph theory as "2-isomorphism" [98, 97]. The networks that can be trivially simplified, namely those which contain a series or parallel connection of the same type of component, are excluded. This enumeration leads to a set of 148 essentially distinct RLC networks with at most five elements of which at most two are reactive.

Of these 148 networks, 40 networks are further eliminated as follows (see Sections 5.3 and 7.1 for the explicit formulae for the Zobel, Cauer–Foster, and Y-Δ transformations):

- Eight networks with four resistors and one reactive element (four with graph structure S or $\mathtt{S}^{\mathrm{d}}$ and four with graph structure V) are eliminated, since their impedance is a bilinear function which can be realized by simpler networks.

- Four networks with four elements (with graph structure G or $\mathtt{G}^{\mathrm{d}}$) can be reduced by a Zobel transformation to the three-element networks #15 and #17.

- Twenty series-parallel networks with five elements can be reduced by a Zobel transformation to networks with four elements or less. Specifically, four networks with graph structure L reduce to networks #20, #25, #28, #32; one with graph structure M reduces to network #72; five with graph structure S reduce to networks #22, #24, #30, #33, #73; five with graph structure $\mathtt{S}^{\mathrm{d}}$ reduce to networks #37, #40, #45, #48, #72; one with graph structure $\mathtt{M}^{\mathrm{d}}$ reduces to network #73; and four with graph structure $\mathtt{L}^{\mathrm{d}}$ reduce to networks #35, #39, #43, #47.

- The four series-parallel networks (with graph structure O and $\mathtt{O}^{\mathrm{d}}$) shown in Figure 5.1, and thence also the four bridge networks, can be reduced to the four-element networks #21, #29, #36, #44 with a Cauer–Foster transformation. For reasons that will become clear in the analysis, this transformation is not considered as a true equivalence (see Section 7.1). However, for each of these networks one of the coefficients A, C, D, or F in (5.1) is zero, and it is straightforward to show that any impedance realizable by such networks can hence also be realized by a network with fewer elements (e.g., by means of the observation that the impedance function is regular—see [47] and Section 7.2).

The 108 networks of the canonical set are shown in Appendix C. The numbering from the Ladenheim catalogue (although not entirely logical) has been preserved. The derivation of the canonical set will be discussed again in Section 7.2.

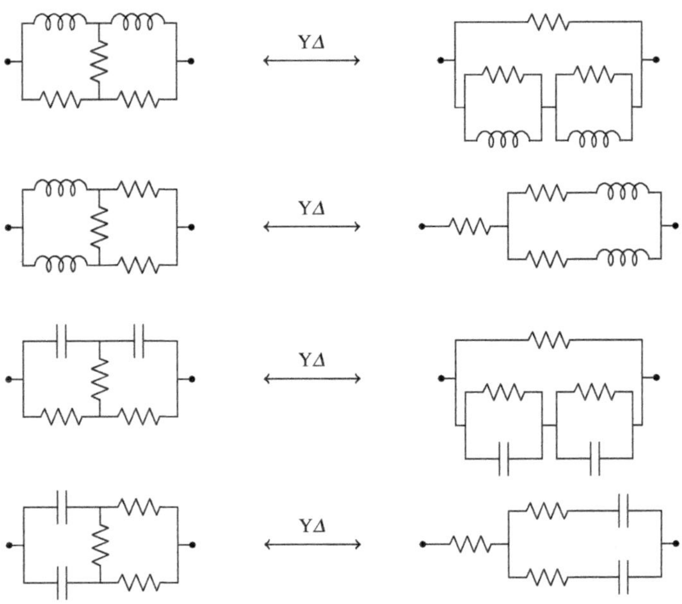

Figure 5.1. *Eight networks which are not part of the catalogue, since they realize impedances which can also be realized by four-element networks.*

Abstractly we can therefore think of the Ladenheim catalogue as a set X containing 108 elements. The individual elements can be defined with varying levels of structure:

1. an oriented graph with two external terminals (the driving-point terminals) in which each branch consists of one of three types of elements (resistor, capacitor, and inductor) with at most three resistors and the sum of the capacitors and inductors being no greater than two;
2. a set of oriented graphs with a fixed structure as described in 1, but with the branch element parameters varying over all real positive numbers.

Here, we are mostly considering an element to have the additional structure defined in 2, namely each element is actually a set. We refer to each element in the catalogue as a *network*, with network numbers according to the Ladenheim enumeration of Appendix C.

5.2 ▪ Approach to classification

To enhance our understanding of the Ladenheim catalogue we seek to uncover as much structure as possible. Our main tools for this purpose, which we describe in the following two sections, are (1) the notion of equivalent network and (2) the use of group actions to formalize certain well-known transformations. The use of these tools in the classification of the networks in the catalogue will be described in Chapter 6.

5.2.1 ▪ Equivalence and realizability set

For each network in the catalogue we are interested in determining the set of impedance functions that a given network can realize when expressed in the form of (5.1). Our first claim is that the set of real numbers that the coefficients A, B, ..., F may assume for a given network is a *semi-algebraic set*. This can be seen as follows. Without loss of generality we will assume $A, B, \ldots, F \geq 0$. For a given network, the impedance $Z(s)$ can be computed as a biquadratic in s,

$$Z(s) = \frac{f_2 s^2 + f_1 s + f_0}{g_2 s^2 + g_1 s + g_0}, \tag{5.2}$$

with coefficients that are polynomial functions of the network parameters R_1, R_2, L_1, etc. Equating (5.2) with a candidate biquadratic impedance (5.1) leads to six polynomial equations of the form $kA = f_2(R_1, R_2, \ldots)$, etc., for some positive constant k. In addition there are (up to) six inequalities: $k > 0$, $R_1 > 0$, etc. Taken together these comprise (up to) six polynomial equations and six polynomial inequalities in the (up to) 12 variables, which define a semi-algebraic set in the 12 parameters (variables). If we project this set onto the first six parameters A, B, ..., F, then we again obtain a semi-algebraic set using the Tarski–Seidenberg theorem [5] which is a subset of $\mathbb{R}^6_+ = \{(x_1, x_2, \ldots, x_6) \text{ s.t. } x_i \geq 0 \text{ for } i = 1, \ldots, 6\}$. We will call this the *realizability set* of the network and we will denote it by $\mathcal{S}_n$, where n is the network number (according to the enumeration in Appendix C).

We define two networks #p and #q to be *equivalent* if $\mathcal{S}_p = \mathcal{S}_q$. This equivalence relation induces a partition of the catalogue into *equivalence classes*. The objective of this monograph is to determine all the semi-algebraic sets $\mathcal{S}_n$ for $n = 1, 2, \ldots, 108$. This allows the complete set of equivalences for the Ladenheim catalogue to be determined, and hence all equivalence classes. For those networks in which it is convenient to use the canonical form for biquadratics described in Section 3.1.2, the semi-algebraic set $\mathcal{S}_n$ can be further reduced to a semi-algebraic set $\mathcal{T}_n$ in the three variables U, V, and W. We note that the realizability region defined in Section 3.1.2, corresponding to the set of realizable impedances for a fixed value of W, is also a semi-algebraic set in the two variables U and V.

More generally we can define the notion of realizability set for an arbitrary two-terminal RLC network. The driving-point impedance of the network takes the form

$$Z(s) = \frac{f_r s^r + f_{r-1} s^{r-1} + \cdots + f_0}{g_r s^r + g_{r-1} s^{r-1} + \cdots + g_0},$$

where $f_r, \ldots, f_0$ and $g_r, \ldots, g_0$ are polynomial functions of the network elements R_1, R_2, L_1, etc., and the integer r cannot exceed the number of reactive elements. Consider a candidate impedance function of the form

$$\frac{a_r s^r + a_{r-1} s^{r-1} + \cdots + a_0}{b_r s^r + b_{r-1} s^{r-1} + \cdots + b_0}.$$

Then by equating the two expressions above, a series of $2(r+1)$ equations of the form $k a_r = f_r(R_1, R_2, \ldots, L_1, L_2, \ldots, C_1, C_2, \ldots)$ results for some positive constant k. Eliminating $k > 0$, $R_1 > 0$, etc., a system of polynomial equations and inequalities in $a_0, \ldots, a_r, b_0, \ldots, b_r$ is obtained which defines a semi-algebraic set $\mathcal{S}$ in $\mathbb{R}^{2r+2}_+$, the realizability set of the network. The sets may also be defined abstractly within $\mathbb{P}^{2r+1}$, the real projective space of dimension $2r+1$. It may sometimes be convenient to embed the realizability set in a higher dimensional space, as we have done for the Ladenheim catalogue, where all realizability sets are considered to belong to $\mathbb{R}^6_+$, even if the number of reactive elements is one or zero. In particular we use the notation of (5.1) for the candidate impedance with $A = D = 0$ when $r = 1$ and $A = B = D = E = 0$ when $r = 0$.

5.2.2 ▪ Group action

The classification of networks is further facilitated by the following transformations on the impedance $Z(s)$:

1. frequency inversion: $s \to s^{-1}$;

2. impedance inversion: $Z \to Z^{-1}$.

As noted in [47], the first transformation corresponds to replacing inductors with capacitors of reciprocal values (and vice versa) and the second to taking the network dual (as described in Section 2.1). We refer to these transformations as $\mathtt{s}$ and $\mathtt{d}$. We further define a transformation which is the composition of the two: $\mathtt{p} = \mathtt{sd}$.

Defining in addition the identity element $\mathtt{e}$, we see that $G = \{\mathtt{e}, \mathtt{s}, \mathtt{d}, \mathtt{p}\}$ is in fact the Klein 4-group, which has the following group table:

	e	d	s	p
e	e	d	s	p
d	d	e	p	s
s	s	p	e	d
p	p	s	d	e

We may then define a *group action* on the set of networks X by $x \to gx$, where $x \in X$ and $g \in G$. This group action induces a partition of X into *orbits* [4]. In our case the orbits comprise one, two, or four elements. In [47] these orbits were referred to as *quartets* and it was noted that sometimes quartets could reduce to two or one element(s).

It was also noted in [47] that frequency inversion corresponds to the transformation $W \leftrightarrow W^{-1}$ in canonical form, and duality corresponds to the transformation $U \leftrightarrow V$, $W \leftrightarrow W^{-1}$. It is easily seen that the transformation $\mathtt{p}$ corresponds to $U \leftrightarrow V$ in canonical form. Therefore, knowing the realizability conditions in canonical form for a given network in an orbit, the derivation of the conditions for the other networks in the orbit is immediate. The notation introduced in Section 3.1.2 is useful in writing the realizability conditions for all the networks in a given orbit, as frequency inversion corresponds to $*$ and the transformation $\mathtt{p}$ corresponds to $\dagger$.

In this monograph we will depart from previous convention by depicting orbits in terms of the two actions $\mathtt{s}$ and $\mathtt{p}$, rather than $\mathtt{s}$ and $\mathtt{d}$. This is in part motivated by the fact that $\mathtt{s}$-invariance can occur independently of $\mathtt{p}$-invariance, while $\mathtt{d}$-invariance always implies $\mathtt{s}$-invariance within the catalogue—a matter that will be studied in more detail in Section 7.5. (We say that a network is $\mathtt{s}$-invariant if the network remains the same after the $\mathtt{s}$ transformation, and similarly with $\mathtt{d}$ and $\mathtt{p}$.) It is also the case that the $\mathtt{p}$ transformation takes a simpler form with respect to the canonical form than the $\mathtt{d}$ transformation.

5.3 ▪ Classical equivalences

We review here two well-known equivalences from linear network analysis. We refer to the first transformation presented as the "Zobel transformation." This appears in explicit form in Zobel [106, Appendix III], together with other transformations, though it is clear from [106] that this transformation was common knowledge at the time. The well-known Y-Δ transformation, which follows, was first published by Kennelly [54].

5.3.1 ▪ Zobel transformation

For any two impedances Z_1 and Z_2, the networks in Figure 5.2 are equivalent in the sense defined in Section 5.2.1 when

$$a' = \frac{a(a+b)}{b}, \qquad b' = a+b, \qquad c' = c\left(\frac{a+b}{b}\right)^2,$$

$$\left[a = \frac{a'b'}{a'+b'}, \qquad b = \frac{(b')^2}{a'+b'}, \qquad c = c'\left(\frac{b'}{a'+b'}\right)^2\right]$$

for any real positive numbers a, b, etc. It is clear from the expressions above that, for any positive and finite value of a, b, and c (respectively, a', b', c') in the transformation, coefficients a', b', and c' (respectively, a, b, c) are necessarily finite and strictly positive.

Figure 5.2. *Zobel transformation.*

5.3.2 ▪ Y-Δ transformation

For any real positive values R_1, R_2, etc., the networks in Figure 5.3 are equivalent when

$$R_1 = \frac{R_b\,R_c}{R_a+R_b+R_c}, \qquad R_2 = \frac{R_a\,R_c}{R_a+R_b+R_c}, \qquad R_3 = \frac{R_a\,R_b}{R_a+R_b+R_c},$$

$$\left[R_a = \frac{R_P}{R_1}, \quad R_b = \frac{R_P}{R_2}, \quad R_c = \frac{R_P}{R_3}, \quad \text{where } R_P = R_1R_2 + R_2R_3 + R_1R_3\right].$$

Figure 5.3. *Y-Δ transformation.*

Chapter 6
Structure of the Ladenheim catalogue

We now proceed to examine the structure of the Ladenheim catalogue using the approach of Chapter 5. It is routine to verify that the group action defined in Section 5.2.2 induces a partition of the catalogue X into 35 orbits. A more difficult task is to identify the further structure that is revealed by the notion of equivalence introduced in Section 5.2.1. Our first step in this regard is to identify all the equivalences that result from two classical transformations, the Zobel transformation and the Y-Δ transformation, described in Section 5.3. This results in a number of orbits "coalescing" through equivalence, and it is convenient to attach a numbering to the resulting "subfamilies," which are 24 in number. Subfamilies are numbered with Roman numerals, according to the number of elements in the networks, with subscript letters to distinguish the subfamilies according to types (e.g., subfamilies of four-element networks are numbered $\mathrm{IV_A}$, $\mathrm{IV_B}$, etc.).

At this point it is unclear whether there are further equivalences which cause some of these subfamilies to further coalesce. This turns out not to be the case with our notion of equivalence (as formalized in Theorem 7.5). To verify this, it is necessary to determine the realizability set $\mathcal{S}_n$ for one representative of each of the 24 subfamilies. This is the main body of work contained in this monograph, the results of which are summarized in Section 6.5. From our analysis it also turns out that some networks, which were classically thought to be equivalent through a Cauer–Foster transformation, are in fact not equivalent (see Section 7.1 for more detail).

Figures 6.1–6.3 in Section 6.1 show the subfamilies and their internal structure consisting of orbits and equivalence classes. These figures summarize the principal structure of the catalogue identified in this monograph. In abstract terms, the 24 subfamilies represent a partition of the catalogue into the "finest common coarsening" of two partitions generated by (i) orbits of the group action and (ii) equivalence classes due to network equivalence. This may be viewed as a main theorem of the monograph whose proof relies on identifying the realizability sets for every subfamily and showing that they are pairwise distinct (Theorem 7.6).

The chapter continues with Sections 6.2, 6.3, and 6.4, which show the mapping and inverse mapping between impedance coefficients and circuit parameters for one-, two-, and three-element networks, four-element networks, and five-element networks, respectively. A characterization of the realizability set for one representative of each subfamily is derived, in terms of necessary and sufficient conditions. Such conditions are summarized in Section 6.5 for all 62 equivalence classes in the catalogue. We note that knowing the realizability set $\mathcal{S}_n$ for a network in a given equivalence class, it can also be easily determined for all other equivalence classes in the same subfamily, by an appropriate transformation of the conditions. More specifically, it is easily shown that the frequency inversion s transformation corresponds to replacing (A, B, C, D, E, F) in the realizability conditions by (C, B, A, F, E, D), while the p transformation corresponds to replacing (A, B, C, D, E, F) by (F, E, D, C, B, A) (the transformations in terms of the canonical form coefficients U, V, and W have already been given in Section 5.2.2). Finally, in Section 6.6, a graphical representation of the realizability region is provided for one equivalence class in each of the five-element subfamilies.

6.1 ▪ Catalogue subfamily structure with orbits and equivalences

A diagrammatic representation of the subfamilies, orbits, and equivalence classes of the catalogue is shown in Figures 6.1, 6.2, and 6.3. Network equivalences are represented through dashed arrows and define the equivalence classes shaded in grey (with one-network equivalence classes not shaded). Equivalence classes are identified by a superscript number (e.g., the two equivalence classes of subfamily $\mathrm{V_G}$ are $\mathrm{V_G^1}$ and $\mathrm{V_G^2}$). Frequency inversion (i.e., s) and the p transformation are indicated through arrows, while duality (i.e., d) and identity (i.e., e) are not shown. Appendix D shows the Ladenheim networks arranged corresponding to the structure of Figures 6.1–6.3.

In Table 6.1 orbits are classified according to the number of networks they contain. Table 6.2 shows the number of equivalence classes, orbits, and networks in all 24 subfamilies. Finally, in Table 6.3 the subfamilies are classified according to the graph topology of the networks they comprise.

Table 6.1. *Classification of orbits according to the number of networks they contain, listed by subfamily.*

Orbits with 1 network	Orbits with 2 networks	Orbits with 4 networks
$\mathrm{I_A}$	$\mathrm{I_B}$	
	$\mathrm{II_B}$	$\mathrm{II_A}$
	$\mathrm{III_D}$, $\mathrm{III_E}$	$\mathrm{III_A}$, $\mathrm{III_B}$, $\mathrm{III_C}$
	$\mathrm{IV_D}$ (×2), $\mathrm{IV_F}$	$\mathrm{IV_A}$ (×3), $\mathrm{IV_B}$, $\mathrm{IV_C}$ (×2), $\mathrm{IV_E}$
$\mathrm{V_I}$	$\mathrm{V_C}$, $\mathrm{V_D}$, $\mathrm{V_F}$, $\mathrm{V_G}$ (×2), $\mathrm{V_H}$	$\mathrm{V_A}$ (×3), $\mathrm{V_B}$ (×2), $\mathrm{V_E}$ (×3), $\mathrm{V_F}$
Tot. 2	Tot. 13	Tot. 20

Table 6.2. *Number of equivalence classes, orbits, and networks in each subfamily.*

	Subfamily	# Eq. classes	# Orbits	# Networks
1-element networks	I_A	1	1	1
	I_B	2	1	2
2-element networks	II_A	4	1	4
	II_B	2	1	2
3-element networks	III_A	2	1	4
	III_B	2	1	4
	III_C	4	1	4
	III_D	2	1	2
	III_E	2	1	2
4-element networks	IV_A	4	3	12
	IV_B	4	1	4
	IV_C	4	2	8
	IV_D	2	2	4
	IV_E	4	1	4
	IV_F	2	1	2
5-element networks	V_A	2	3	12
	V_B	4	2	8
	V_C	2	1	2
	V_D	2	1	2
	V_E	4	3	12
	V_F	2	2	6
	V_G	2	2	4
	V_H	2	1	2
	V_I	1	1	1
Total	24	62	35	108

Table 6.3. *Classification of the subfamilies according to the type of networks they contain. Simple series-parallel networks are denoted by SSP and series-parallel networks by SP (see definitions in Section* 2.1*).*

Subfamily	Network type
1-, 2-, 3-element subfamilies	SSP
IV_A, IV_C, IV_D, IV_E	SSP
IV_B, IV_F	SP
V_A, V_E	SSP
V_B, V_F	SP
V_C, V_D, V_H, V_I	Bridge
V_G	SP / Bridge

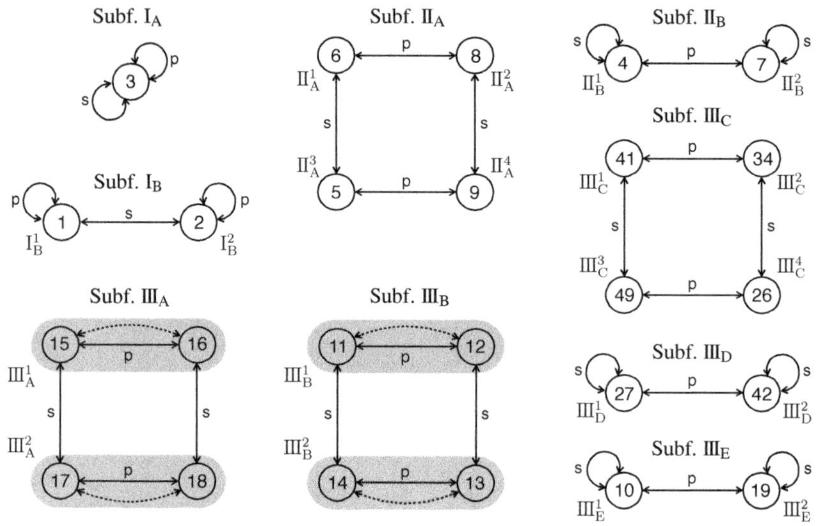

Figure 6.1. *One-, two-, and three-element subfamilies, orbits, and equivalence classes. All equivalences (dashed arrows) are the Zobel transformation defined in Section* 5.3. *One-network equivalence classes are not shaded.*

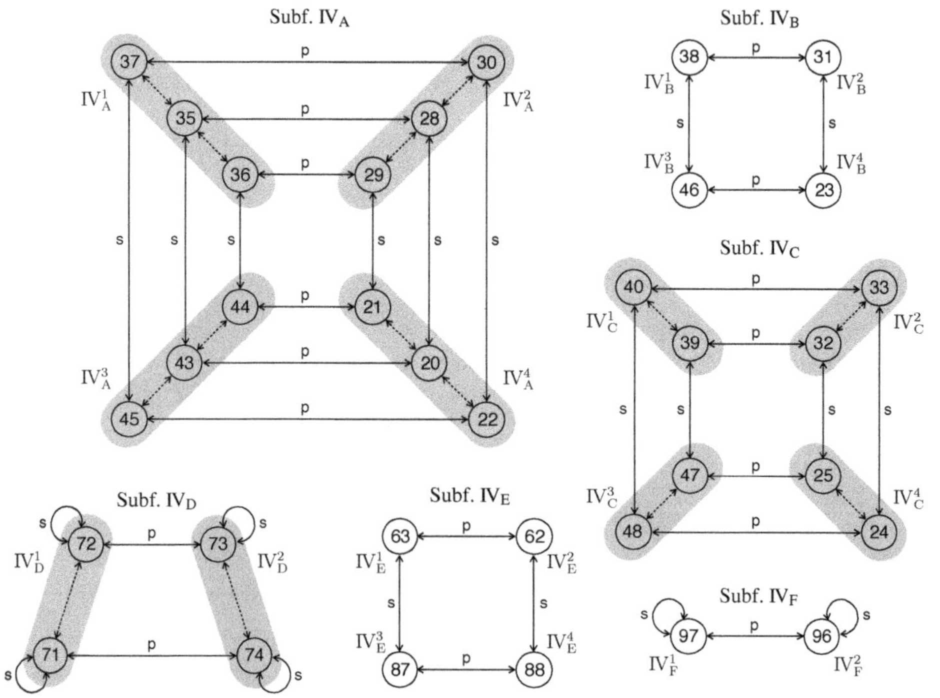

Figure 6.2. *Four-element subfamilies, orbits, and equivalence classes. All equivalences (dashed arrows) are the Zobel transformation defined in Section* 5.3. *One-network equivalence classes are not shaded.*

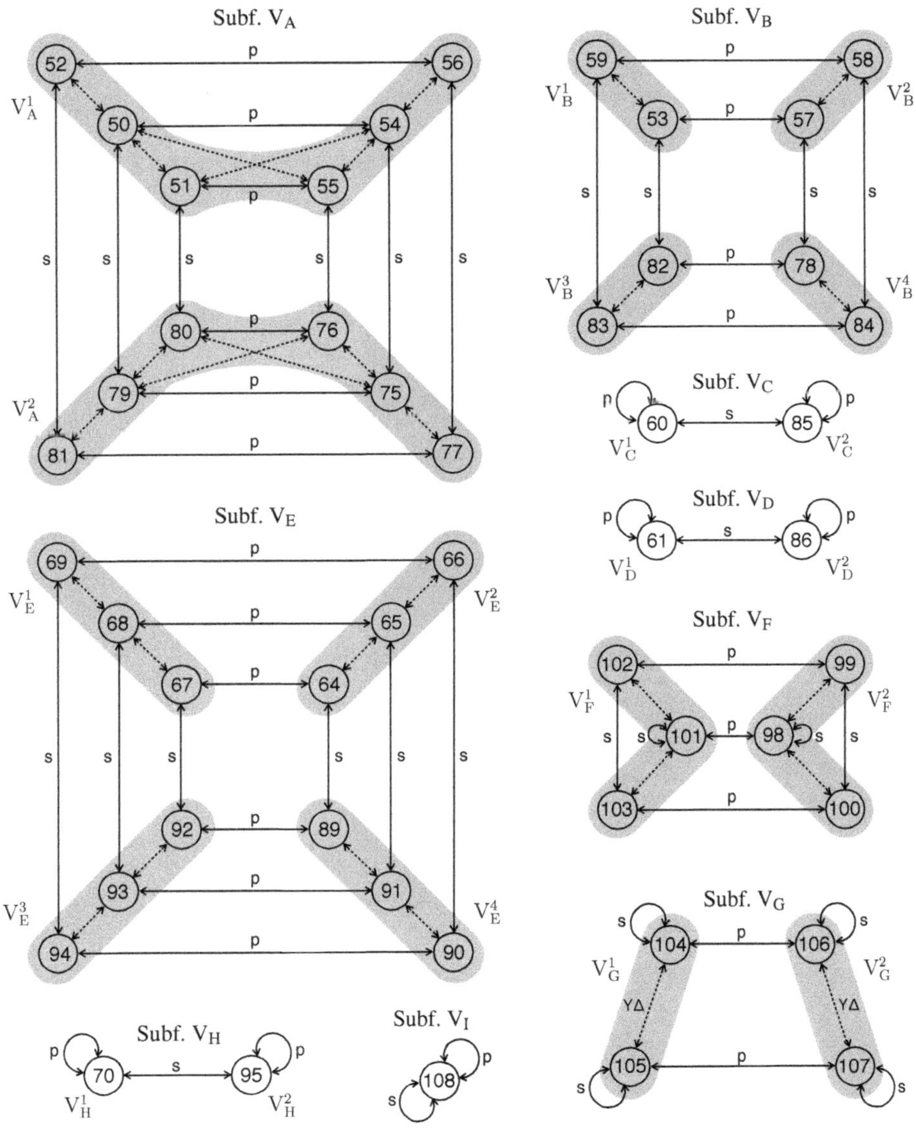

Figure 6.3. *Five-element subfamilies, orbits, and equivalence classes. Unless indicated otherwise, all equivalences (dashed arrows) are the Zobel transformation defined in Section* 5.3. *One-network equivalence classes are not shaded.*

6.2 ▪ One-, two-, and three-element networks

The derivation of the realizability set for a given network begins with the formulae for the impedance coefficients in terms of the network parameters, from which the inverse mapping may be studied. Table 6.4 contains expressions for coefficients $A, B, \ldots, F$ for one representative network in each subfamily of one-, two-, and three-element networks. Such expressions can be easily found by computing the impedance of each network. Table 6.5 shows the result of the inverse problem; i.e., expressions for the inductances, capacitances, and resistances in each network are shown in terms of coefficients A, $B, \ldots, F$. The inverse problem is quite straightforward for networks with at most three elements, and hence the elimination procedure which leads to the expressions in Table 6.5 is not shown.

Table 6.4. *Expressions for $A, B, \ldots, F$ in terms of resistances, inductances, and capacitances for one-, two-, and three-element networks.*

Network No.	Equiv. class	A	B	C	D	E	F
3	I^1_A	0	0	R_1	0	0	1
1	I^1_B	0	L_1	0	0	0	1
6	II^1_A	0	L_1	R_1	0	0	1
4	II^1_B	L_1C_1	0	1	0	C_1	0
15	$\mathrm{III}^1_\mathrm{A}$	0	$L_1(R_1+R_2)$	R_1R_2	0	L_1	R_1
11	$\mathrm{III}^1_\mathrm{B}$	L_1L_2	$R_1(L_1+L_2)$	0	0	L_1	R_1
41	$\mathrm{III}^1_\mathrm{C}$	$R_1L_1C_1$	L_1	0	L_1C_1	R_1C_1	1
27	$\mathrm{III}^1_\mathrm{D}$	$R_1L_1C_1$	L_1	R_1	L_1C_1	0	1
10	$\mathrm{III}^1_\mathrm{E}$	L_1C_1	R_1C_1	1	0	C_1	0

Realizability conditions are summarized in Section 6.5 for the representative equivalence classes of one-, two-, and three-element networks. These conditions are easily deduced from the requirement of positivity of the expressions in Tables 6.4 and 6.5. Realizability conditions for all other equivalence classes in the subfamilies can be easily found by an appropriate transformation of the polynomials appearing in the conditions, as described at the beginning of this chapter.

6.3 ▪ Four-element networks

The analysis of four-element networks is slightly more complicated than the case with three or fewer elements. Table 6.6 shows the expressions for coefficients $A, B, \ldots, F$ for one network in each subfamily of four-element networks, which again can be easily found by computing the impedance of the network. Table 6.7 contains expressions for inductances, capacitances, and resistances for each network in terms of coefficients A, $B, \ldots, F$. For all subfamilies other than $\mathrm{IV_B}$ the derivation of such expressions in explicit form is quite straightforward.

Table 6.5. *Expressions for inductances, capacitances, and resistances in terms of coefficients A, B, ..., F for one-, two-, and three-element networks.*

Network No.	Equiv. class	R_1	R_2	L_1	L_2	C_1
3	$\mathrm{I}_{\mathrm{A}}^{1}$	C/F	–	–	–	–
1	$\mathrm{I}_{\mathrm{B}}^{1}$	–	–	B/F	–	–
6	$\mathrm{II}_{\mathrm{A}}^{1}$	C/F	–	B/F	–	–
4	$\mathrm{II}_{\mathrm{B}}^{1}$	–	–	A/E	–	E/C
15	$\mathrm{III}_{\mathrm{A}}^{1}$	$\dfrac{BF-CE}{EF}$	C/F	$\dfrac{BF-CE}{F^2}$	–	–
11	$\mathrm{III}_{\mathrm{B}}^{1}$	$\dfrac{-(AF-BE)}{E^2}$	–	$\dfrac{-(AF-BE)}{EF}$	A/E	–
41	$\mathrm{III}_{\mathrm{C}}^{1}$	A/D	–	B/F	–	D/B
27	$\mathrm{III}_{\mathrm{D}}^{1}$	C/F	–	B/F	–	D/B
10	$\mathrm{III}_{\mathrm{E}}^{1}$	B/E	–	A/E	–	E/C

The following remarks should be made regarding the derivation of the realizability conditions given in Table 6.7 for four-element networks:

- For network #37 in equivalence class $\mathrm{IV}_{\mathrm{A}}^{1}$ and for network #40 in equivalence class $\mathrm{IV}_{\mathrm{C}}^{1}$, it is easily seen that

$$K|_{C=0} = F\big[A(AF-BE)+B^2D\big]\,, \tag{6.1}$$

  where the expression for K can be found in Table 6.10. From (6.1), $K<0$ implies $AF-BE<0$, and hence the condition on $AF-BE$ can be omitted from the realizability conditions for equivalence class $\mathrm{IV}_{\mathrm{A}}^{1}$ if the condition on K is included.

- For network #72 in equivalence class $\mathrm{IV}_{\mathrm{D}}^{1}$, condition $AF-CD=0$ follows from the expressions for A, F, C, and D in Table 6.6 and implies that $R_2=A/D=C/F$.

- For network #63 in equivalence class $\mathrm{IV}_{\mathrm{E}}^{1}$, the condition $\lambda_1=0$ follows from the expressions for $A, B, \ldots, F$ in Table 6.6.

- For network #97 in equivalence class $\mathrm{IV}_{\mathrm{F}}^{1}$, τ_1 is defined in Table 6.10, and condition $\tau_1=0$ once again follows from the expressions for $A, B, \ldots, F$ in Table 6.6.

The realizability conditions summarized in Section 6.5 follow. For subfamily IV_{B}, a more thorough analysis of the realizability conditions is needed, and Theorem A.1 provides a full characterization of the realizability set.

Table 6.6. *Expressions for A, B, …, F in terms of resistances, inductances, and capacitances for four-element networks.*

Network	A	B	C	D	E	F
#37 ($\mathrm{IV}_{\mathrm{A}}^{1}$)	$L_1L_2(R_1+R_2)$	$L_2R_1R_2$	0	L_1L_2	$L_1(R_1+R_2)+L_2R_1$	R_1R_2
#38 ($\mathrm{IV}_{\mathrm{B}}^{1}$)	$L_1L_2(R_1+R_2)$	$R_1R_2(L_1+L_2)$	0	L_1L_2	$L_1R_2+L_2R_1$	R_1R_2
#40 ($\mathrm{IV}_{\mathrm{C}}^{1}$)	$R_1R_2L_1C_1$	$L_1(R_1+R_2)$	0	$R_1L_1C_1$	$L_1+R_1R_2C_1$	R_1+R_2
#72 ($\mathrm{IV}_{\mathrm{D}}^{1}$)	$R_2L_1C_1$	$R_1R_2C_1$	R_2	L_1C_1	$(R_1+R_2)C_1$	1
#63 ($\mathrm{IV}_{\mathrm{E}}^{1}$)	$L_1C_1(R_1+R_2)$	$L_1+R_1R_2C_1$	R_2	L_1C_1	R_1C_1	1
#97 ($\mathrm{IV}_{\mathrm{F}}^{1}$)	$R_1L_1C_1$	$R_1R_2C_1+L_1$	R_2	L_1C_1	$(R_1+R_2)C_1$	1

Table 6.7. *Expressions for inductances, capacitances, and resistances in terms of coefficients A, B, …, F for four-element networks.*

Network	R_1	R_2	L_1	L_2	C_1
#37 ($\mathrm{IV}_{\mathrm{A}}^{1}$)	$\dfrac{K}{DF(AF-BE)}$	$\dfrac{-B^2}{AF-BE}$	$\dfrac{-BK}{F(AF-BE)^2}$	B/F	–
#38 ($\mathrm{IV}_{\mathrm{B}}^{1}$)	See Theorem A.1				
#40 ($\mathrm{IV}_{\mathrm{C}}^{1}$)	$\dfrac{-K}{DF(AF-BE)}$	A/D	B/F	–	$\dfrac{BD^2F}{K}$
#72 ($\mathrm{IV}_{\mathrm{D}}^{1}$)	$\dfrac{-BC}{BF-CE}$	A/D	$\dfrac{-DC^2}{F(BF-CE)}$	–	$\dfrac{-(BF-CE)}{C^2}$
#63 ($\mathrm{IV}_{\mathrm{E}}^{1}$)	$\dfrac{AF-CD}{DF}$	C/F	$\dfrac{AF-CD}{EF}$	–	$\dfrac{DE}{AF-CD}$
#97 ($\mathrm{IV}_{\mathrm{F}}^{1}$)	A/D	C/F	$\dfrac{AF+CD}{EF}$	–	$\dfrac{DE}{AF+CD}$

6.4 ▪ Five-element networks

We finally consider the most interesting case of five-element networks. Table 6.8 shows the expressions for coefficients A, B, …, F for one representative network in each of the nine five-element subfamilies, which can be found by computing the impedance of the network (see, e.g., [74, Section 7.2]). Table 6.9 contains expressions for inductances, capacitances, and resistances for subfamilies V_{A} and V_{E}; for all the other subfamilies, expressions for the network elements can be found in the theorems referenced in the table.

The characterization of the realizability set $\mathcal{S}_n$ for each representative network in terms of necessary and sufficient conditions has been derived in this program of work in Theorems A.2–A.10. We note that realizability conditions for bridge networks in

Table 6.8. *Expressions for A, B, …, F in terms of resistances, inductances, and capacitances for five-element networks.*

Network	A	B	C
#52 (V_A^1)	$L_1L_2(R_1+R_2+R_3)$	$L_1R_3(R_1+R_2)+L_2R_1(R_2+R_3)$	$R_1R_2R_3$
#59 (V_B^1)	$L_1L_2(R_2+R_3)$	$L_1R_2R_3+L_2(R_1R_2+R_1R_3+R_2R_3)$	$R_1R_2R_3$
#60 (V_C^1)	$L_1L_2(R_1+R_3)$	$L_2(R_1R_2+R_1R_3+R_2R_3)+R_2L_1(R_1+R_3)$	$R_1R_2R_3$
#61 (V_D^1)	$L_1L_2(R_1+R_2+R_3)$	$L_1R_1(R_2+R_3)+L_2R_2(R_1+R_3)$	$R_1R_2R_3$
#69 (V_E^1)	$R_1L_1C_1(R_2+R_3)$	$L_1(R_1+R_2+R_3)+R_1R_2R_3C_1$	$R_3(R_1+R_2)$
#101 (V_F^1)	$L_1C_1R_1R_3$	$R_3(R_1R_2C_1+L_1)$	R_2R_3
#104 (V_G^1)	$L_1C_1(R_1+R_3)$	$L_1+C_1(R_1R_2+R_1R_3+R_2R_3)$	R_2+R_3
#70 (V_H^1)	$L_1C_1(R_1R_2+R_1R_3+R_2R_3)$	$R_1R_2R_3C_1+L_1(R_1+R_3)$	$R_2(R_1+R_3)$
#108 (V_H^1)	$L_1C_1R_3(R_1+R_2)$	$C_1R_1R_2R_3+L_1(R_1+R_2+R_3)$	$R_1(R_2+R_3)$

Network	D	E	F
#52 (V_A^1)	L_1L_2	$L_2R_1+L_1(R_1+R_2)$	R_1R_2
#59 (V_B^1)	L_1L_2	$L_1R_2+L_2(R_1+R_3)$	$R_2(R_1+R_3)$
#60 (V_C^1)	L_1L_2	$L_2(R_1+R_2)+L_1(R_1+R_2+R_3)$	$R_3(R_1+R_2)$
#61 (V_D^1)	L_1L_2	$L_1(R_2+R_3)+L_2(R_1+R_3)$	$R_1R_2+R_1R_3+R_2R_3$
#69 (V_E^1)	$R_1L_1C_1$	$L_1+R_1R_2C_1$	R_1+R_2
#101 (V_F^1)	$L_1C_1(R_1+R_3)$	$C_1(R_1R_2+R_1R_3+R_2R_3)+L_1$	R_2+R_3
#104 (V_G^1)	L_1C_1	$C_1(R_1+R_2)$	1
#70 (V_H^1)	$L_1C_1(R_2+R_3)$	$C_1R_1(R_2+R_3)+L_1$	$R_1+R_2+R_3$
#108 (V_H^1)	$L_1C_1(R_1+R_2)$	$C_1(R_1R_2+R_1R_3+R_2R_3)+L_1$	R_2+R_3

Table 6.9. *Expressions for inductances, capacitances, and resistances in terms of coefficients $A, B, \ldots, F$ for five-element networks.*

Network	R_1	R_2	R_3	L_1	L_2	C_1
#52 (V_A^1)	$\frac{-K}{D\lambda_1}$	$\frac{(BF-CE)^2}{F\lambda_1}$	$\frac{C}{F}$	$\frac{-K(BF-CE)}{\lambda_1^2}$	$\frac{BF-CE}{F^2}$	-
#59 (V_B^1)	See Theorem A.3					
#60 (V_C^1)	See Theorem A.4					
#61 (V_D^1)	See Theorem A.5					
#69 (V_E^1)	$\frac{K}{D\lambda_1}$	$\frac{AF-CD}{DF}$	$\frac{C}{F}$	$\frac{BF-CE}{F^2}$	-	$\frac{D^2(BF-CE)}{K}$
#101 (V_F^1)	See Theorem A.7					
#104 (V_G^1)	See Theorem A.8					
#70 (V_H^1)	See Theorem A.9					
#108 (V_I^1)	See Theorem A.10					

subfamilies V_C, V_D, V_G, and V_I have also been independently derived in Chen et al. in 2016 [18, Theorems 1, 2, 4, 5], and for subfamilies V_A, V_E, and V_H in [47]. In [18], the multiplicity of solutions is not taken into account, and some conditions are expressed in a different form. Some considerations on the smallest generating set are made in Section 7.3 in relation to results found in [18].

Table 6.10 lists the expressions for all the polynomials which appear in the realizability conditions, both in terms of the biquadratic impedance coefficients $A, B, \ldots, F$ of (5.1) and in terms of the coefficients U, V, and W of the canonical form (3.12). The notation for the polynomials in canonical form is described in Section 3.1.2.

6.5 ▪ Summary of realizability conditions

Table 6.11 summarizes the realizability conditions for all equivalence classes in the catalogue. Expressions for the symbols appearing in the conditions can be found in Table 6.10. Unless indicated otherwise, we assume $A, B, \ldots, F > 0$. The notation regarding the number of solutions has the following meaning:

- 1/2 Depending on the orbit, there can be one or two solutions.
- ∞ There are infinitely many solutions, since one of the network elements can take any value within a certain interval (while the other elements can be computed according to the formulae in the realization theorems).
- * When any of the quantities in the conditions is zero, there is only one solution.

In networks with one, two, or three elements, the cases in which $C = F = 0$ have not been considered, since they would lead to a trivial cancellation of the frequency variable s at the numerator and denominator.

Table 6.10. *Polynomials appearing in the realizability conditions, expressed in terms of both A, B, ..., F and U, V, W. The expressions in U, V, W are obtained by replacing A, B, C, ... by 1, $2U\sqrt{W}$, W, ... (from (3.12)), except for a multiplicative positive scaling.*

Special polynomials in terms of $A, B, \ldots, F$	Reduced expressions in terms of U, V, W
$K = (AF-CD)^2-(AE-BD)(BF-CE)$	$K_c = 4(U^2+V^2) - 4UV(W^{-1}+W)$ $+(W^{-1}-W)^2$
$\sigma = BE - (\sqrt{AF} - \sqrt{CD})^2$	$\sigma_c = 4UV + 2 - (W^{-1} + W)$
$\lambda_1 = E(BF - CE) - F(AF - CD)$	$\lambda_c = 4UV - 4V^2W - (W^{-1} - W)$
$\lambda_2 = B(AE - BD) - A(AF - CD)$	$\lambda_c^\dagger = 4UV - 4U^2W - (W^{-1} - W)$
$\lambda_3 = D(AF - CD) - E(AE - BD)$	$\lambda_c^* = 4UV - 4V^2W^{-1} - (W - W^{-1})$
$\lambda_4 = C(AF - CD) - B(BF - CE)$	$\lambda_c^{*\dagger} = 4UV - 4U^2W^{-1} - (W - W^{-1})$
$\eta = (AF+CD)^2-(AE-BD)(BF-CE)$	$\eta_c = 4(U^2+V^2) - 4UV(W+W^{-1})$ $+(W^{-1}+W)^2$
$\mu_1 = K - 4CD(2AF - 2CD - BE)$	$\mu_c = 4(U^2+V^2) + 4UV(3W - W^{-1})$ $+(1-W^{-2})(9W^2-1)$
$\mu_2 = K - 4AF(2CD - 2AF - BE)$	$\mu_c^* = 4(U^2+V^2) + 4UV(3W^{-1} - W)$ $+(1-W^2)(9W^{-2}-1)$
$\tau_1 = K - DF(B^2 - 4AC)$	$\tau_c = 4V^2 - 4UV(W^{-1}+W)$ $+(W^{-1}+W)^2$
$\tau_2 = K - AC(E^2 - 4DF)$	$\tau_c^\dagger = 4U^2 - 4UV(W^{-1}+W)$ $+(W^{-1}+W)^2$
$\delta = BE - 2(AF + CD)$	$\delta_c = 4UV - 2(W^{-1} + W)$
$\zeta_1 = -E(BF - CE) + 2F(AF - CD)$	$\zeta_c = 4V(V - UW^{-1}) + 2(W^{-2} - 1)$
$\zeta_2 = -B(AE - BD) + 2A(AF - CD)$	$\zeta_c^\dagger = 4U(U - VW^{-1}) + 2(W^{-2} - 1)$
$\zeta_3 = E(AE - BD) - 2D(AF - CD)$	$\zeta_c^* = 4V(V - UW) + 2(W^2 - 1)$
$\zeta_4 = B(BF - CE) - 2C(AF - CD)$	$\zeta_c^{*\dagger} = 4U(U - VW) + 2(W^2 - 1)$
$\psi = (AF + CD)(K + 4ACDF)$ $-2ABCDEF$	$\psi_c = 4(W^{-1}+W)(U^2+V^2) -$ $4UV(W^2+4+W^{-2}) + (W^{-1}+W)^3$
$\rho_1 = -K + 2CD(AF - CD)$	$\rho_c = -4(U^2+V^2) + 4UV(W+W^{-1})$ $-(1-W^{-2})(3W^2-1)$
$\rho_2 = -K + 2AF(CD - AF)$	$\rho_c^* = -4(U^2+V^2) + 4UV(W+W^{-1})$ $-(1-W^2)(3W^{-2}-1)$
$AF - CD$	$W^{-1} - W$
$E^2 - 4DF$	$4W^{-1}(V^2 - 1)$
$B^2 - 4AC$	$4W(U^2 - 1)$

Table 6.11. *Realizability conditions and number of solutions for all equivalence classes in the catalogue.*

Subf.	Eq. class	Networks	Realizability conditions	# sol.
I_A	I_A^1	3	$A = B = D = E = 0$	1
I_B	I_B^1	1	$A = C = D = E = 0$	1
	I_B^2	2	$A = B = D = F = 0$	1
II_A	II_A^1	6	$A = D = E = 0$	1
	II_A^2	8	$A = C = D = 0$	1
	II_A^3	5	$A = D = F = 0$	1
	II_A^4	9	$A = B = D = 0$	1
II_B	II_B^1	4	$B = D = F = 0$	1
	II_B^2	7	$A = C = E = 0$	1
III_A	III_A^1	15, 16	$A = D = 0, BF - CE > 0$	1
	III_A^2	17, 18	$A = D = 0, BF - CE < 0$	1
III_B	III_B^1	11, 12	$C = D = 0, AF - BE < 0$	1
	III_B^2	13, 14	$A = F = 0, BE - CD > 0$	1
III_C	III_C^1	41	$C = 0, AF - BE = 0$	1
	III_C^2	34	$D = 0, AF - BE = 0$	1
	III_C^3	49	$A = 0, BE - CD = 0$	1
	III_C^4	26	$F = 0, BE - CD = 0$	1
III_D	III_D^1	27	$E = 0, AF - CD = 0$	1
	III_D^2	42	$B = 0, AF - CD = 0$	1
III_E	III_E^1	10	$D = F = 0$	1
	III_E^2	19	$A = C = 0$	1
IV_A	IV_A^1	35, 36, 37	$K < 0, C = 0$	1
	IV_A^2	28, 29, 30	$K < 0, D = 0$	1
	IV_A^3	43, 44, 45	$K < 0, A = 0$	1
	IV_A^4	20, 21, 22	$K < 0, F = 0$	1

Table 6.11 (cont.)

Subf.	Eq. class	Networks	Realizability conditions	# sol.
IV_B	IV_B^1	38	1. $C = 0, K < 0$ 2. $C = 0, K = 0, E^2 - 4DF = 0$	1 ∞
	IV_B^2	31	1. $D = 0, K < 0$ 2. $D = 0, K = 0, B^2 - 4AC = 0$	1 ∞
	IV_B^3	46	1. $A = 0, K < 0$ 2. $A = 0, K = 0, E^2 - 4DF = 0$	1 ∞
	IV_B^4	23	1. $F = 0, K < 0$ 2. $F = 0, K = 0, B^2 - 4AC = 0$	1 ∞
IV_C	IV_C^1	39, 40	$K > 0, C = 0, AF - BE < 0$	1
	IV_C^2	32, 33	$K > 0, D = 0, AF - BE < 0$	1
	IV_C^3	47, 48	$K > 0, A = 0, BE - CD > 0$	1
	IV_C^4	24, 25	$K > 0, F = 0, BE - CD > 0$	1
IV_D	IV_D^1	71, 72	$AF - CD = 0, BF - CE < 0$	1
	IV_D^2	73, 74	$AF - CD = 0, BF - CE > 0$	1
IV_E	IV_E^1	63	$\lambda_1 = 0, AF - CD > 0$	1
	IV_E^2	62	$\lambda_2 = 0, AF - CD > 0$	1
	IV_E^3	87	$\lambda_3 = 0, AF - CD < 0$	1
	IV_E^4	88	$\lambda_4 = 0, AF - CD < 0$	1
IV_F	IV_F^1	97	$\tau_1 = 0$	1
	IV_F^2	96	$\tau_2 = 0$	1
V_A	V_A^1	50, 51, 52, 54, 55, 56	$K < 0, AF - CD > 0$	1
	V_A^2	75, 76, 77, 79, 80, 81	$K < 0, AF - CD < 0$	1
V_B	V_B^1	53, 59	1. $AF - CD > 0, K < 0$ 2. $AF - CD > 0, K = 0, E^2 - 4DF = 0$	1/2 ∞
	V_B^2	57, 58	1. $AF - CD > 0, K < 0$ 2. $AF - CD > 0, K = 0, B^2 - 4AC = 0$	1/2 ∞
	V_B^3	82, 83	1. $AF - CD < 0, K < 0$ 2. $AF - CD < 0, K = 0, E^2 - 4DF = 0$	1/2 ∞
	V_B^4	78, 84	1. $AF - CD < 0, K < 0$ 2. $AF - CD < 0, K = 0, B^2 - 4AC = 0$	1/2 ∞

Table 6.11 (cont.)

Subf.	Eq. class	Networks	Realizability conditions	# sol.
V_C	V_C^1	60	$\eta \leq 0$, $AF - CD > 0$	2^*
	V_C^2	85	$\eta \leq 0$, $AF - CD < 0$	2^*
V_D	V_D^1	61	$K \leq 0$, $\mu_1 \leq 0$, $AF - CD > 0$	2^*
	V_D^2	86	$K \leq 0$, $\mu_2 \leq 0$, $AF - CD < 0$	2^*
V_E	V_E^1	67, 68 ,69	$K > 0$, $\lambda_1 > 0$, $AF - CD > 0$	1
	V_E^2	64, 65, 66	$K > 0$, $\lambda_2 > 0$, $AF - CD > 0$	1
	V_E^3	92, 93, 94	$K > 0$, $\lambda_3 > 0$, $AF - CD < 0$	1
	V_E^4	89, 90, 91	$K > 0$, $\lambda_4 > 0$, $AF - CD < 0$	1
V_F	V_F^1	101, 102, 103	1. $K > 0$, $\tau_1 < 0$	1
			2. $K = \tau_1 = 0$	∞
	V_F^2	98, 99, 100	1. $K > 0$, $\tau_2 < 0$	1
			2. $K = \tau_2 = 0$	∞
V_G	V_G^1	104, 105	1. $AF - CD > 0$	
			a. $\tau_1 < 0$, $\lambda_1 = 0$	1
			b. $\lambda_1 > 0$, $\tau_1 = 0$, $\delta > 0$	1
			c. $\tau_1 < 0$, $\lambda_1 > 0$, $K \geq 0$, $\delta > 0$, $\zeta_1 > 0$	2^*
			2. $AF - CD \geq 0$, $\tau_1\lambda_1 > 0$	1
			3. $AF - CD = 0$, $K = 0$	∞
			4. $AF - CD < 0$ and one of	
			a. $\tau_1 < 0$, $\lambda_3 = 0$	1
			b. $\lambda_3 > 0$, $\tau_1 = 0$, $\delta > 0$	1
			c. $\tau_1 < 0$, $\lambda_3 > 0$, $K \geq 0$, $\delta > 0$, $\zeta_3 > 0$	2^*
			5. $AF - CD \leq 0$, $\tau_1\lambda_3 > 0$	1
V_G	V_G^2	106, 107	1. $AF - CD > 0$	
			a. $\tau_2 < 0$, $\lambda_2 = 0$	1
			b. $\lambda_2 > 0$, $\tau_2 = 0$, $\delta > 0$	1
			c. $\tau_2 < 0$, $\lambda_2 > 0$, $K \geq 0$, $\delta > 0$, $\zeta_2 > 0$	2^*
			2. $AF - CD \geq 0$, $\tau_2\lambda_2 > 0$	1
			3. $AF - CD = 0$, $K = 0$	∞
			4. $AF - CD < 0$ and one of	
			a. $\tau_2 < 0$, $\lambda_4 = 0$	1
			b. $\lambda_4 > 0$, $\tau_2 = 0$, $\delta > 0$	1
			c. $\tau_2 < 0$, $\lambda_4 > 0$, $K \geq 0$, $\delta > 0$, $\zeta_4 > 0$	2^*
			5. $AF - CD \leq 0$, $\tau_2\lambda_4 > 0$	1

Table 6.11 (cont.)

Subf.	Eq. class	Networks	Realizability conditions	# sol.
V_H	V_H^1	70	$\mu_1 \geq 0$, $AF - CD > 0$ and one of	
			1. signs of λ_1, λ_2, ρ_1 not all the same	2^*
			2. $\lambda_1 = \lambda_2 = 0$, $\rho_1 = 0$	∞
	V_H^2	95	$\mu_2 \geq 0$, $AF - CD < 0$ and one of	
			1. signs of λ_3, λ_4, ρ_2 not all the same	2^*
			2. $\lambda_3 = \lambda_4 = 0$, $\rho_2 = 0$	∞
V_I	V_I	108	$K \geq 0$ and one of	
			1. $\tau_1\tau_2 < 0$	1
			2. $\tau_1 = 0$, $\tau_2 < 0$, $\psi > 0$	1
			3. $\tau_2 = 0$, $\tau_1 < 0$, $\psi > 0$	1
			4. $\tau_1 < 0$, $\tau_2 < 0$, $\psi > 0$	2^*
			5. $\tau_1 = \tau_2 = 0$, $\psi = 0$	∞

6.5.1 ▪ Realization procedure for a biquadratic impedance

We illustrate here a possible approach for the synthesis of a candidate biquadratic impedance which makes use of the information summarized in Table 6.11. Given an impedance in the form (5.1), the first step is to verify whether it is realizable by a network in the Ladenheim catalogue. We recall that networks in the catalogue can realize all the regular biquadratics and a subset of the nonregular biquadratics (namely the nonregular impedances realized by equivalence classes V_H^1 and V_H^2). Using the results summarized in Section 3.1.1 we can easily check whether the impedance is regular, from which these three cases follow:

1. The impedance is regular, hence realizable by one or more networks in the catalogue. By computing some of the polynomial quantities which appear in the realizability conditions in Table 6.11 (e.g., K, $AF - CD$, τ_1, τ_2, etc.) it is possible to find all the equivalence classes within the catalogue that realize the given impedance, as well as the multiplicity of the solutions.

2. The impedance is nonregular, but the realizability conditions of either V_H^1 or V_H^2 hold, and hence the impedance is realizable by either network #70 or network #95.

3. The impedance is nonregular and not realizable by a network in subfamily V_H. A realization of the impedance with five elements or less is therefore not possible, and an alternative synthesis method among those described in Chapters 2 and 3 must be used, for example, a possible realization with six elements as in [49] or the Bott–Duffin method.

6.6 ▪ Realizability regions for five-element networks

We conclude this chapter by showing a graphical representation of the realizability region for one equivalence class in each of the nine five-element subfamilies. The realizability regions are obtained from the conditions summarized in Table 6.11 expressed in canonical form (see Table 6.10 for expressions for all the polynomials appearing in the realizability conditions in terms of U, V, and W). The regions are then plotted in the (U, V)-plane for significant values of W, as shown in Figures 6.4–6.12 (hatched regions). We recall that in every figure the grey region corresponds to $\sigma_c < 0$ and represents nonpositive-real biquadratics, whereas the region corresponding to $\lambda_c > 0$ and/or $\lambda_c^\dagger > 0$ represents regular biquadratics for $W \leq 1$: it can be seen that all but one subfamily (i.e., subfamily $\mathrm{V_H}$) realize regular biquadratics, as pointed out in [47].

Figures 6.13 and 6.14 show the number of distinct networks which can realize impedances in a given region of the (U, V)-plane, again for significant values of W, as well as the equivalence classes to which such networks belong. If a network can realize a given impedance with two distinct combinations of values of resistances, inductances, and capacitances, we consider that there are two distinct solutions and the network is counted twice.

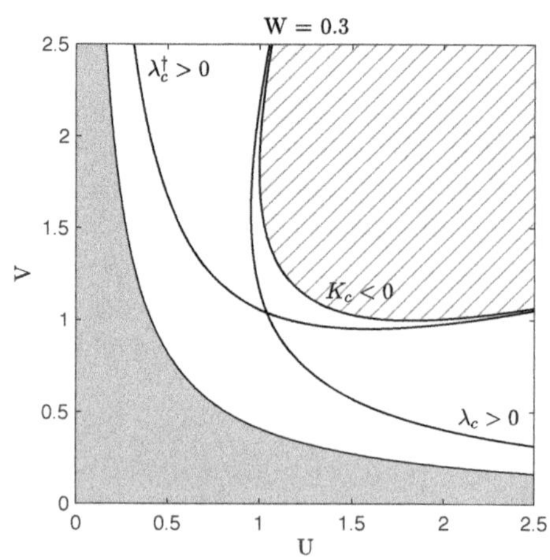

Figure 6.4. *Equivalence class V_A^1. The interior of the hatched region is realizable, with multiplicity of solution equal to one.*

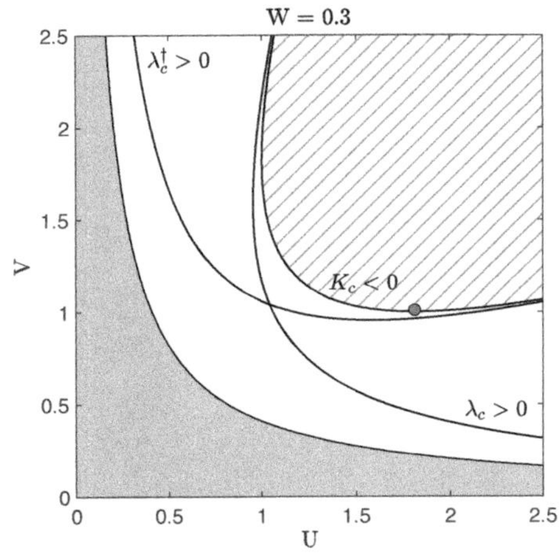

Figure 6.5. *Equivalence class V_B^1. We note that the realizability region is the same as equivalence class V_A^1, with the addition of the point on the boundary of K_c in which $V = 1$ (i.e., $E^2 - 4DF = 0$), which corresponds to infinitely many solutions. In the interior of the realizability region there are always two solutions, which may coincide or not depending on which orbit is considered.*

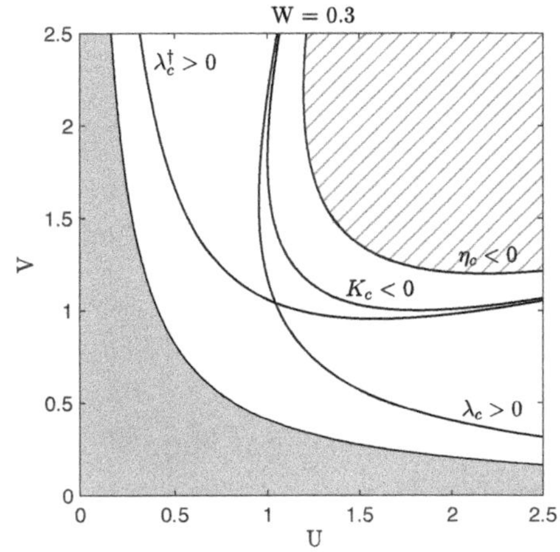

Figure 6.6. *Equivalence class V_C^1. There are always two solutions for this subfamily, which coincide on the boundary of the realizability region.*

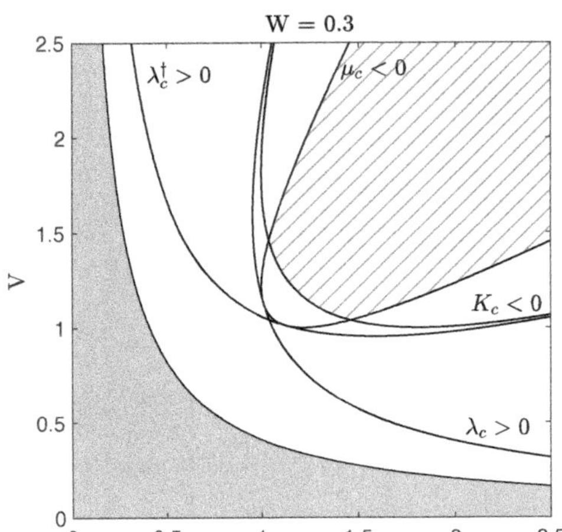

Figure 6.7. *Equivalence class V_D^1. There are always two solutions for this subfamily, which coincide on the boundary of the realizability region.*

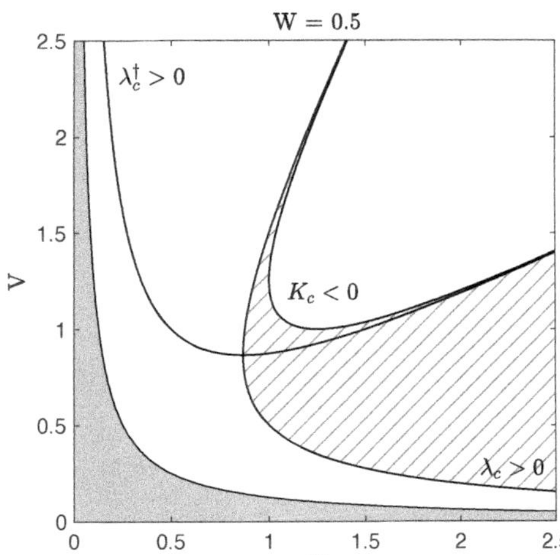

Figure 6.8. *Equivalence class V_E^1. The interior of the hatched region is realizable with multiplicity of solution equal to one.*

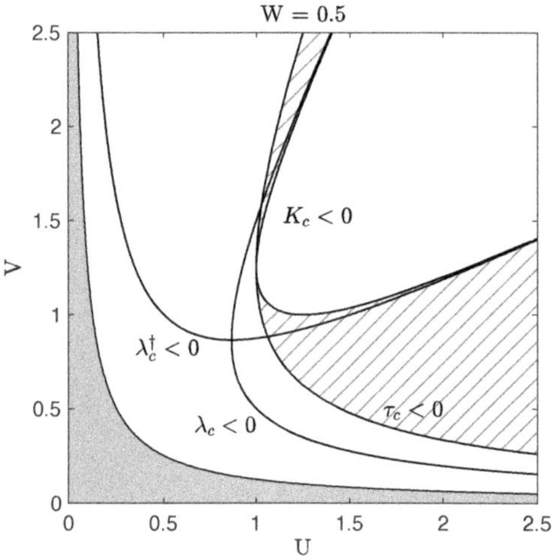

Figure 6.9. *Equivalence class V_F^1. The realizability region is given by the interior of the hatched region, where there is one solution, with the addition of the point where $K_c = \tau_c$, which corresponds to infinitely many solutions.*

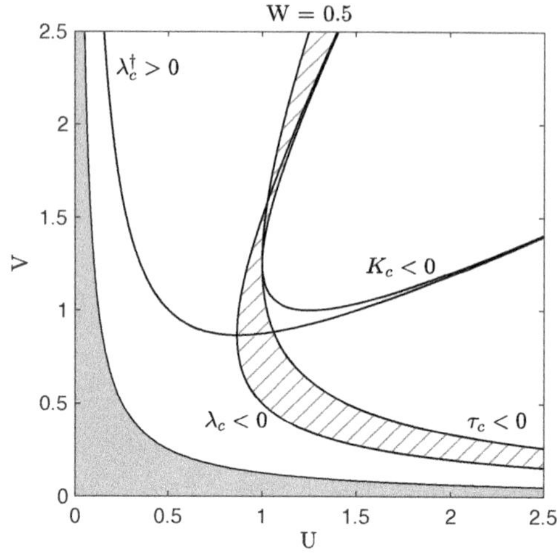

Figure 6.10. *Equivalence class V_G^1. Only the active boundaries for the realizability region have been plotted. See Figure* A.10 *for more details on the boundaries and on the multiplicity of solutions.*

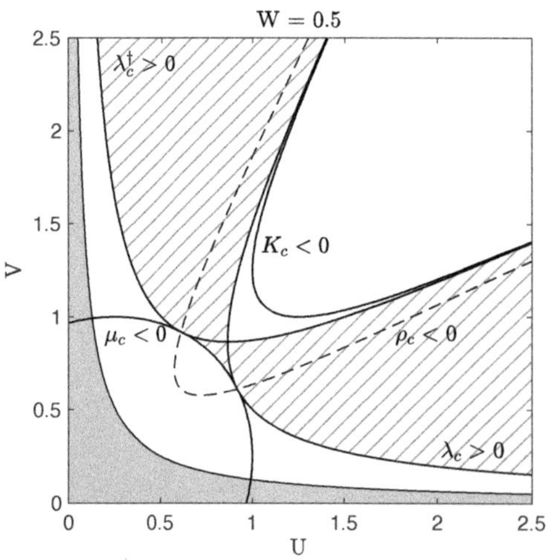

Figure 6.11. *Equivalence class V_H^1. See Figures* A.12 *and* A.13 *and Theorem* A.9 *for more details on the boundaries and on the multiplicity of solutions.*

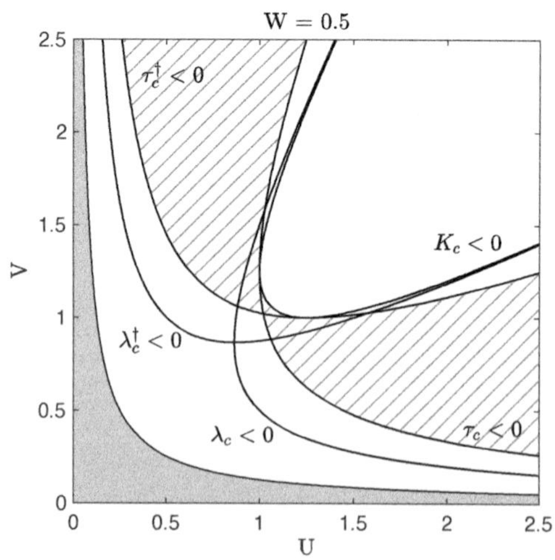

Figure 6.12. *Subfamily V_I. Only the active boundaries for the realizability region have been plotted. See Figure* A.15 *for more details on the boundaries and on the multiplicity of solutions.*

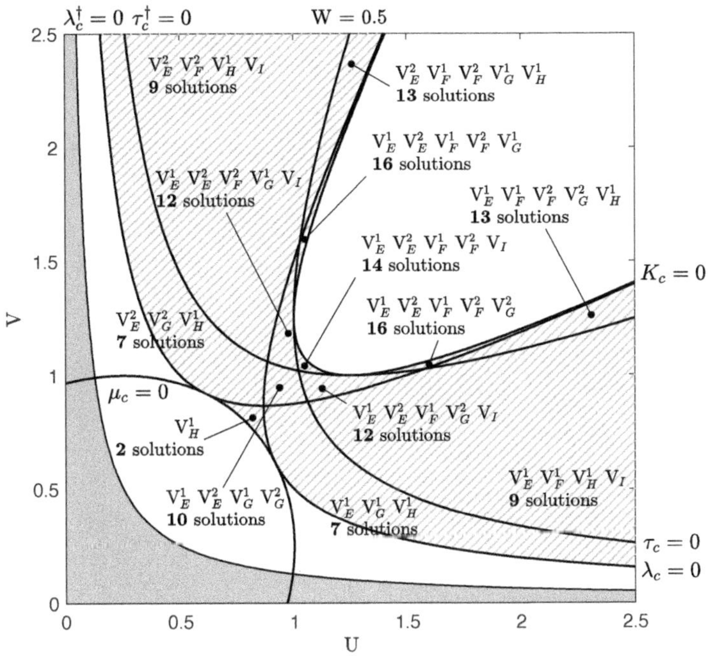

Figure 6.13. *Number of distinct networks (and name of the corresponding equivalence classes) which can realize impedances in all realizable regions with $K_c > 0$ for $W = 0.5$.*

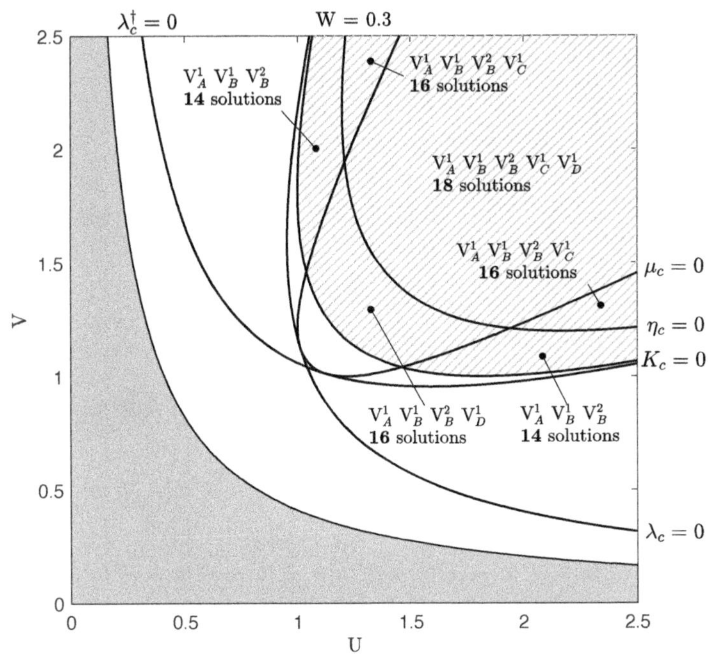

Figure 6.14. *Number of distinct networks (and name of the corresponding equivalence classes) which can realize impedances in all realizable regions with $K_c < 0$ for $W = 0.3$.*

Chapter 7

Main results and discussion

Following the realizability analysis for the 108 networks of the Ladenheim catalogue of Chapter 6 we are ready to assess the structure that has been revealed in the catalogue. Our first task is to consider in more depth the classical Cauer and Foster canonical forms in Section 7.1. We then develop formally some of the main results of the monograph in Section 7.2. A discussion is then given on the smallest set of networks needed to realize any regular biquadratic in Section 7.3, and some remarks are made on conjectures contained in Kalman's latest work in Section 7.4. We conclude the chapter by studying in Section 7.5 properties of invariance to duality in RLC networks and presenting in Section 7.6 two new equivalences which were found by analyzing the class of six-element networks with four resistors.

7.1 ▪ Cauer–Foster transformation

Below are the transformations between the so-called Cauer canonical form and Foster first and second canonical forms (cf. Section 2.2 and [10, 25]). By applying the Zobel transformation of Figure 5.2 to the networks on the left-hand side of Figures 7.1 and 7.2, one can define a number of additional "quasi-equivalences" (a concept that will become clearer at the end of this section). A number of these quasi-equivalences can also be found in [106, Appendix III].

The networks in Figure 7.1 are related by the following transformation:

$$a' = \frac{ac(b+d)^2}{ad^2+b^2c}, \quad b' = b+d, \quad c' = \frac{(ad-bc)^2}{ad^2+b^2c}, \quad d' = \frac{bd(b+d)(ad-bc)^2}{(ad^2+b^2c)^2},$$

$$\left[a = \frac{(b'c')^2-(a'd'+c'd'-M)^2}{4d'M}, \quad b = \frac{-b'(a'd'+c'd'-b'c'-M)}{2M}, \right.$$

$$c = \frac{-(b'c')^2+(a'd'+c'd'+M)^2}{4d'M}, \quad d = \frac{b'(a'd'+c'd'-b'c'+M)}{2M},$$

$$\left. \text{where } M = \sqrt{(a'd'-b'c')^2+c'd'(2a'd'+2b'c'+c'd')} \right]$$

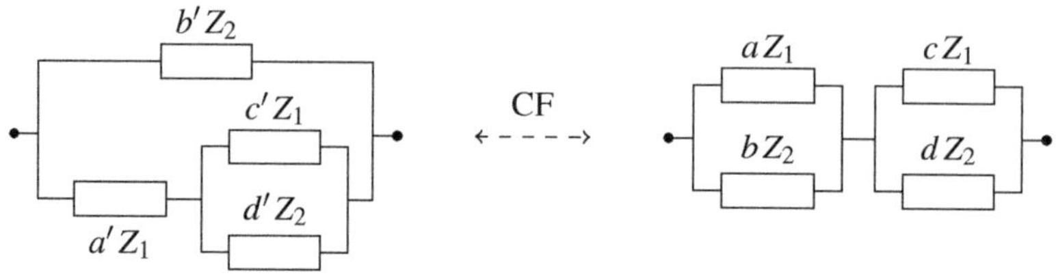

Figure 7.1. *Cauer–Foster quasi-equivalence.*

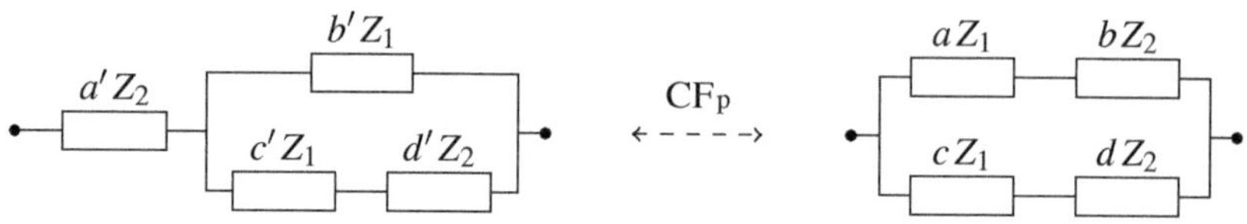

Figure 7.2. p *transformation of Cauer–Foster quasi-equivalence.*

for any real positive numbers a, b, etc., where $Z_1(s)$ and $Z_2(s)$ are arbitrary impedances. The networks in Figure 7.2 are related by the following transformation:

$$a' = \frac{bd}{b+d}, \quad b' = \frac{ad^2+b^2c}{(b+d)^2}, \quad c' = \frac{ac(ad^2+b^2c)}{(ad-bc)^2}, \quad d' = \frac{(ad^2+b^2c)^2}{(b+d)(ad-bc)^2},$$

$$\left[\begin{array}{ll} a = \dfrac{(a'b'+a'c'+b'd'+N)\,N}{d'(-a'b'-a'c'+b'd'+N)}, & b = \dfrac{2a'\,N}{-a'b'-a'c'+b'd'+N}, \\[2ex] c = \dfrac{(a'b'+a'c'+b'd'-N)\,N}{d'(a'b'+a'c'-b'd'+N)}, & d = \dfrac{2a'\,N}{a'b'+a'c'-b'd'+N}, \\[2ex] \text{where } N = \sqrt{(a'b'+a'c')^2+b'd'(2a'b'-2a'c'+b'd')} & \end{array}\right]$$

for any real positive numbers a, b, etc., where $Z_1(s)$ and $Z_2(s)$ are arbitrary impedances. From the formulae for the transformation in Figure 7.1 it follows that if $ad - bc = 0$, with a, b, c, and d finite and positive, then $c' = d' = 0$, and the network reduces to a two-block structure (i.e., the parallel connection of $a'Z_1$ and $b'Z_2$). This means that the network on the right-hand side can be reduced to a two-block structure with a suitable choice of *strictly positive* and finite coefficients a, b, c, and d, while a similar reduction for the network on the left-hand side requires the coefficients c' and d' to be zero. Similar considerations hold for the transformation in Figure 7.2, which is the dual form of the transformation in Figure 7.1, in the sense of duality of the graph but not of the network elements.

Subfamilies IV_A–IV_B and V_A–V_B (which are, respectively, Cauer forms and Foster forms) are related by the transformations above, as shown in Figures 7.3 and 7.4. Classically, the networks of these two pairs of subfamilies were considered to be equivalent. However, the derivation of the realizability conditions for such networks (which are summarized in Section 6.5) led to the following observations:

- For the Cauer forms, the resultant of the numerator and denominator K is strictly negative, while in the Foster forms we can have $K = 0$ (i.e., a pole-zero cancellation).

- There is only one solution to the realization problem for the Cauer forms, while there are two solutions for the Foster forms. These solutions are identical for the networks in subfamily $\mathrm{IV_B}$ and in the inner orbit of subfamily $\mathrm{V_B}$, due to the symmetry of such networks, while for the networks in the outer orbit of $\mathrm{V_B}$ the two solutions are distinct.

The two forms therefore define different realizability sets (as also can be seen from Figures 6.4 and 6.5), and hence the networks are not truly equivalent according to our definition. We can define a "weaker" type of equivalence, or *quasi-equivalence*, compared to the transformations of Section 5.3 if the realizability sets of two networks differ only on a subset of lower dimension (cf. Definition 7.1). We also note that, by considering the networks of the above-mentioned four subfamilies (which are related through the Zobel transformation and p transformation/frequency inversion as illustrated in Figures 7.3–7.4), we obtain the complete set of quasi-equivalences for the catalogue.

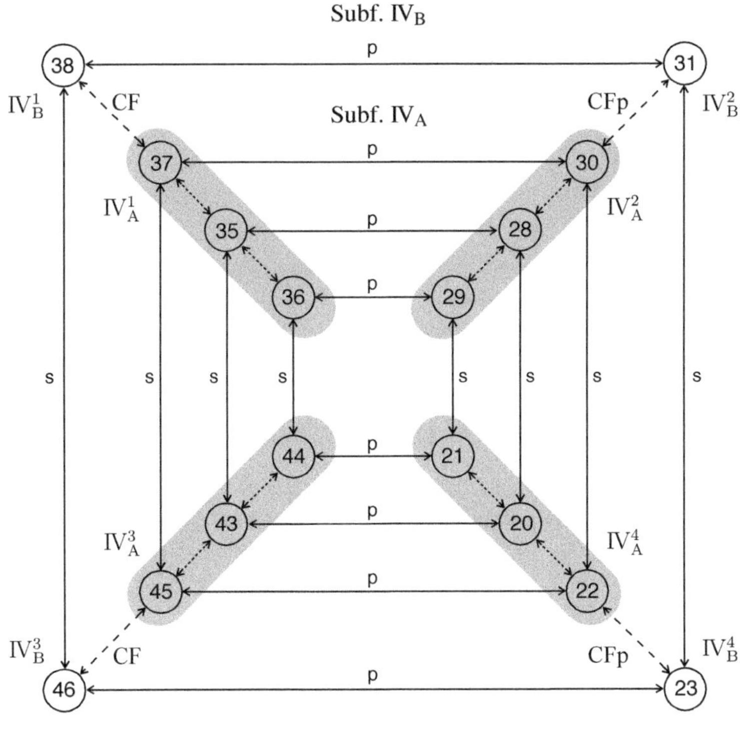

Figure 7.3. *Relation between subfamilies IV_A–IV_B through Cauer–Foster transformations. CF indicates the Cauer–Foster transformation of Figure* 7.1, *while CFp indicates the transformation of Figure* 7.2.

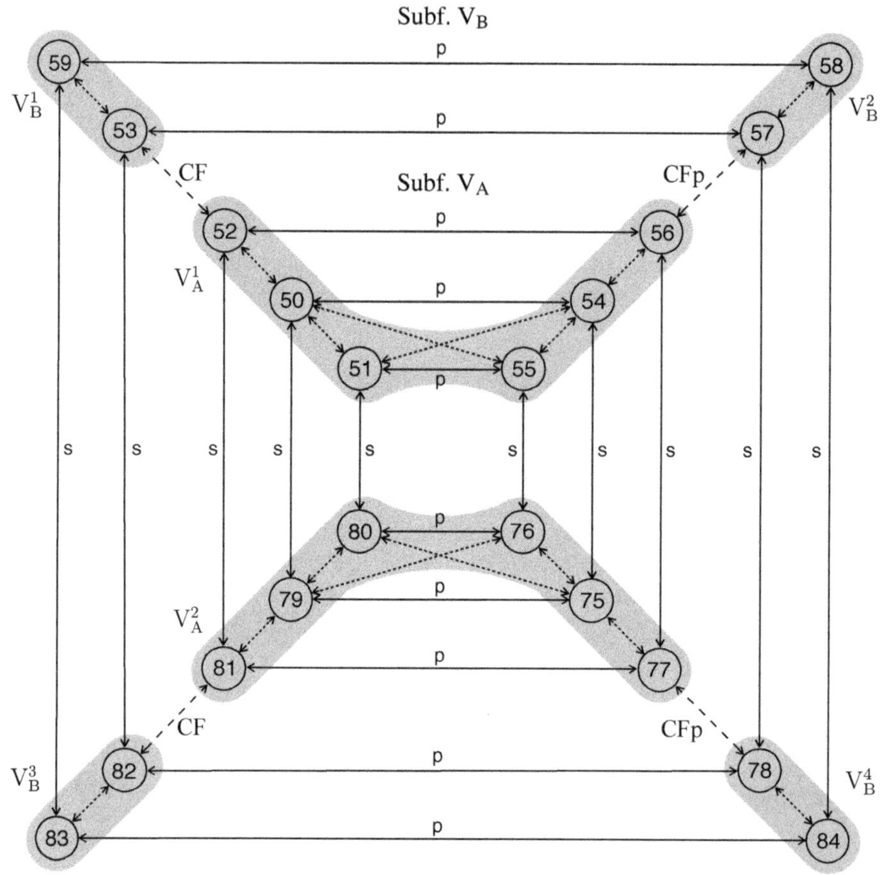

Figure 7.4. *Relation between subfamilies V_A–V_B through Cauer–Foster transformations. CF indicates the Cauer–Foster transformation of Figure 7.1, while CFp indicates the transformation of Figure 7.2.*

7.2 ▪ Formal results on the Ladenheim catalogue

Table 6.11 in Section 6.5 summarizes the realizability conditions for the networks of the catalogue as polynomial equations and inequalities in the variables A, B, …, F. These implicitly define the semi-algebraic sets $\mathcal{S}_n$ of Section 5.2.1 as subsets of $\mathbb{R}^6_+$. We state a standard definition for the dimension of a semi-algebraic set [3] and propose a definition of a generic network.

Definition 7.1. *The* dimension $\dim(\mathcal{S})$ *of a semi-algebraic set* $\mathcal{S}$ *is defined as the largest d such that there exists a one-to-one smooth map from the open cube* $(-1,1)^d \subset \mathbb{R}^d$ *into* $\mathcal{S}$.

Definition 7.2. *An RLC two-terminal network containing* m *elements is* generic *if* $\dim(\mathcal{S}) = m + 1$, *where* $\mathcal{S}$ *is the realizability set of the network.*

Theorem 7.3. *All* 108 *networks of the Ladenheim catalogue are generic.*

Proof. We begin with the five-element networks. For each network we observe that there exists a point $(A_0, B_0, \dots, F_0) \in \mathcal{S}_n$ and $\bar{\epsilon} > 0$ such that $(A_0 + \epsilon_1, B_0 + \epsilon_2, \dots, F_0 + \epsilon_6) \in \mathcal{S}_n$ providing $|\epsilon_i| < \bar{\epsilon}$ for $i = 1, \dots, 6$. (All that is required is to find a point whose feasibility is determined only by polynomial strict inequalities. For example, for $n = 104$, it is sufficient to find a point satisfying 2 in the corresponding entry in Table 6.11 for which $AF - CD > 0$.) For such a point we have the mapping

$$(x_1, x_2, \dots, x_6) \in (-1,1)^6 \to (A_0 + x_1\bar{\epsilon}, B_0 + x_2\bar{\epsilon}, \dots, F_0 + x_6\bar{\epsilon}) \in \mathcal{S}_n,$$

which is one-to-one, and hence the network is generic.

For the four-element networks we will take as an example network #71 for which the realizability conditions are $AF - CD = 0, BF - CE < 0$. Consider any point $(A_0, B_0, \dots, F_0) \in \mathcal{S}_{71}$ and let $A = A_0 + \epsilon_1$, $B = B_0 + \epsilon_2, \dots, E = E_0 + \epsilon_5$, $F = CD/A$. Then there exists an $\bar{\epsilon} > 0$ such that $(A, B, \dots, F) \in \mathcal{S}_{71}$ providing $|\epsilon_i| < \bar{\epsilon}$ for $i = 1, \dots, 5$. Now define a mapping

$$(x_1, x_2, \dots, x_5) \in (-1,1)^5 \to (A_0 + x_1\bar{\epsilon}, \dots, E_0 + x_5\bar{\epsilon}, F_1) \in \mathcal{S}_{71},$$

where $F_1 = (C_0 + x_3\bar{\epsilon})(D_0 + x_4\bar{\epsilon})/(A_0 + x_1\bar{\epsilon})$. This mapping is one-to-one and onto a neighborhood of $(A_0, B_0, \dots, F_0) \in \mathcal{S}_{71}$. We have thus found an open neighborhood of a general point in $\mathcal{S}_{71}$ which is homeomorphic (having a bicontinuous invertible mapping) to the open cube $(-1,1)^5$, and hence the unit sphere, in $\mathbb{R}^5$. It is not possible for such a neighborhood to be homeomorphic to a sphere of different dimension [35, Theorem 2.26], and hence the network is generic. The proof for other four-element networks is similar.

For the three-element networks we will take as an example network #31 for which the realizability conditions are $C = 0, AF - BE = 0$. Consider any point

$$(A_0, B_0, 0, D_0, E_0, F_0) \in \mathcal{S}_{31}$$

and let $A = A_0 + \epsilon_1$, $B = B_0 + \epsilon_2$, $C = 0$, $D = D_0 + \epsilon_3$, $E = E_0 + \epsilon_4$, $F = BE/A$. Then there exists an $\bar{\epsilon} > 0$ such that $(A, B, \dots, F) \in \mathcal{S}_{31}$ providing $|\epsilon_i| < \bar{\epsilon}$ for $i = 1, \dots, 4$. Now define a mapping

$$(x_1, x_2, x_3, x_4) \in (-1,1)^4 \to (A_0 + x_1\bar{\epsilon}, B_0 + x_2\bar{\epsilon}, 0, D_0 + x_3\bar{\epsilon}, E_0 + x_4\bar{\epsilon}, F_1) \in \mathcal{S}_{31},$$

where $F_1 = (B_0 + x_2\bar{\epsilon})(E_0 + x_4\bar{\epsilon})/(A_0 + x_1\bar{\epsilon})$. This mapping is one-to-one and onto a neighborhood of $(A_0, B_0, \dots, F_0) \in \mathcal{S}_{31}$. We have thus found an open neighborhood of a general point in $\mathcal{S}_{31}$ which is homeomorphic to the open cube $(-1,1)^4$, and hence the network is generic. The proof for other three-element networks is similar. For the two- and one-element networks the proof is elementary. □

Theorem 7.4. *The* 108 *networks of the Ladenheim catalogue form the complete set of all essentially distinct (up to* 2*-isomorphism), generic, two-terminal RLC networks comprising at most two reactive elements.*

Proof. The enumeration procedure to determine the 148 essentially distinct networks is as described in Section 5.1. The 40 networks that were eliminated to produce the canonical set are now easily seen to be nongeneric: eight networks with four resistors and one reactive element have a realizability set of dimension 4; four networks with four elements which can be reduced by a Zobel transformation to three-element networks have a realizability set of dimension 4; 20 SP networks with five elements which can be reduced by a Zobel transformation to networks with four elements or less have a realizability set of dimension at most 5; and finally, the eight networks shown in Figure 5.1 have one of the coefficients A, C, D, or F in (5.1) equal to zero and hence have a realizability set of dimension at most 5. □

Theorem 7.5. *The Ladenheim catalogue comprises* 62 *equivalence classes as listed in Table* 6.11 *according to the definition of equivalence in Section* 5.2.1.

Proof. The networks within each class in Table 6.11 have already been seen to be equivalent by a Zobel or Y-Δ transformation. It remains to show that each class is distinct, i.e., the corresponding realizability sets are distinct. From Theorem 7.3 we can immediately conclude that networks with a different number of elements are distinct, since they define realizability sets of different dimension. To complete the proof we must show that any two equivalence classes of networks with the same number of elements are distinct, i.e., there is a point in one realizability set that is not in the other. This is easily seen for equivalence classes within the same subfamily and is trivial for the one-, two-, and three-element cases. For any pair of five-element equivalence classes it is straightforward to find points in one realizability region which are not in the other through the graphical representations in Figures 6.4–6.12 in Section 6.6. Finally, for the four-element networks, we first observe that the equivalence classes of subfamilies IV_A and IV_B (for which $K < 0$) are necessarily distinct from the remaining subfamilies (for which $K > 0$, which is implied by the realizability conditions); subfamilies IV_A and IV_B are distinct from each other, since there is always an extra point in the realizability set for the latter; in IV_C one coefficient is always zero, and hence it is different from IV_D, IV_E, and IV_F; IV_D is necessarily distinct from IV_E (due to the condition on $AF - CD$) and IV_F (since we can have $\tau_1 = 0$ or $\tau_2 = 0$ with $AF - CD \neq 0$ in the latter); and finally, by plotting the corresponding curves of the realizability region, we easily find distinct points in the realizability sets for IV_E and IV_F. □

It should be remarked that the above theorem shows that, within the Ladenheim catalogue, there are no new equivalences among the circuits that are not derived through the Zobel or Y-Δ transformations.

We now show that the 24 subfamilies of the Ladenheim catalogue comprise the *finest common coarsening* of the partitions induced by the equivalence relations of (1) group action and (2) network equivalence. This is also known as the *join* of the two equivalence relations [6, Definition 7, Section 18]. We will be content to state this formally in terms of the notion of transitive closure of a relation. A relation on a set X is a subset R of $X \times X$, and when $\langle a, b \rangle \in R$ we write aRb and say that a and b are related by R. The union $T = R \cup S$ of two relations R and S on X is the union of the corresponding subsets of $X \times X$, hence aTb if and only if aRb or aSb (we note that the union of two equivalence relations is not in general an equivalence relation). Similarly, we say that S contains R and write $R \subseteq S$ if and only if $aRb \Rightarrow aSb$ for all $a, b \in X$ [71, pp. 573–581]. The *transitive closure* of a relation R is the smallest transitive relation containing R.

Theorem 7.6. *The* 24 *subfamilies of the Ladenheim catalogue comprise the partition induced by the transitive closure of the union of the two equivalence relations given by* (1) *group action and* (2) *network equivalence.*

Proof. The two equivalence relations (which we will refer to as R_1 and R_2) generate two partitions of the catalogue X, π_1 and π_2, into 35 orbits and 62 equivalence classes, as shown in Figures 6.1–6.3. The 24 subfamilies also form a partition π of the catalogue, which is generated by an equivalence relation W. If we define the relation $T = R_1 \cup R_2$, we observe that within each block of π there is a finite path in T between any two networks. Therefore W satisfies the property of being the "connectivity relation" of $R_1 \cup R_2$ [71, p. 600], which is the same as the transitive closure of $R_1 \cup R_2$ [71, Section 9.4, Theorem 2]. □

7.3 ▪ Smallest generating set of the catalogue

In [47] it was shown that only two networks (one from each of the two equivalence classes of subfamily $\mathrm{V_A}$) are needed to realize any positive-real biquadratic impedance with resultant $K \leq 0$, while four networks (one from each equivalence class of subfamily $\mathrm{V_E}$) are needed to realize all regular biquadratics with $K \geq 0$. (See (3.3) for the expression for the resultant K.) These six networks, taken together with the only two networks of the catalogue which can realize nonregular impedances (i.e., the networks in subfamily $\mathrm{V_H}$), represent a *generating set* for the Ladenheim class: any impedance which can be realized by a network in the catalogue can also be realized by one of these eight networks. It should be pointed out that the case $K = 0$ involves some element values being taken to be zero or infinity. If one maintains the condition that all element values are finite and nonzero, then the $K = 0$ cases should be covered by an appropriate set of simpler networks with fewer elements.

In [18, Corollary 1] it is shown that the number of networks required to realize all regular biquadratics with $K > 0$ can be reduced by one compared to [47] if one considers subfamily $\mathrm{V_I}$ and one network from each of the two equivalence classes of subfamily $\mathrm{V_G}$ (i.e., three networks in total). Hence, the generating set can be reduced by one compared to [47], at the expense of covering the following special cases with four-element networks, since biquadratics which satisfy the following conditions cannot be realized by networks in subfamilies $\mathrm{V_G}$ and $\mathrm{V_I}$:

1. either λ_1 or λ_2 negative and one of τ_1 or τ_2 zero, (7.1)
2. either λ_3 or λ_4 negative and one of τ_1 or τ_2 zero. (7.2)

(See Table 6.10 for the expressions for the polynomials appearing in the conditions.) The conditions in (7.1) can be deduced by combining the realizability regions of equivalence classes $\mathrm{V_G^1}$, $\mathrm{V_G^2}$, and $\mathrm{V_I}$ (plotted in Figures 6.10 and 6.12 for $AF - CD > 0$), and similarly for (7.2) (when $AF - CD < 0$). We note that (7.1) and (7.2) are boundary cases for the individual networks but end up being in the interior of the region which is realizable by the whole generating set, as one can see from Figure 6.13. As in [47] certain other boundary cases such as where $K = 0$ also require some element values to be zero or infinity, and these also would need to be covered by simpler networks if one maintains the condition that all element values are finite and nonzero.

We remark that the new generating set identified in [18], despite having one fewer network, is not particularly useful in practice, since it introduces more bridge networks and SP networks, as opposed to the generating set in [47] of mostly SSP networks which are easier to realize.

7.4 ▪ Remarks on Kalman's 2011 Berkeley seminar

In this section we will review some of the concepts found in the notes on a talk given by Kalman in Berkeley (October 26, 2011) on electrical network synthesis [59] and draw a connection with the results obtained in our analysis of the Ladenheim catalogue.

The following definitions can be found in [59]. Given a network $\Gamma \in X$ defined by an undirected connected graph G_Γ, the impedance of the network is expressed as

$$Z_\Gamma = \frac{a_\Gamma(s)}{b_\Gamma(s)},$$

where $a_\Gamma(s)$ and $b_\Gamma(s)$ are relatively prime polynomials in s, with degrees α_Γ and β_Γ, respectively. The following polynomial map is defined, which takes the network parameters to the impedance Z_Γ:

$$\psi_\Gamma : \mathbb{R}^r \to \mathbb{P}^{\alpha_\Gamma+\beta_\Gamma+1}.$$

The domain of the map ψ_Γ is the space of parameters for the network (where r is the number of network elements), while the codomain is the projective space $\mathbb{P}^{\alpha_\Gamma+\beta_\Gamma+1}$ derived from the coefficients of the numerator $a_\Gamma(s)$ and denominator $b_\Gamma(s)$. Abstractly, the two problems of analysis and synthesis/realization can be defined through the map ψ_Γ and the inverse map ψ_Γ^{-1}.

In [59], a network Γ is defined to be *minimal* (or *generic*) if the map ψ_Γ is finite-to-one. We have seen that, for networks in the catalogue, when the map ψ_Γ is finite-to-one it can be either one-to-one or two-to-one. In the latter case, there are two distinct solutions to the synthesis problem; i.e., given a network $\Gamma \in X$, one can find two distinct combinations of edge weights (network parameters) which lead to the same impedance. Also, given a network Γ, the multiplicity of solutions can vary with the image point, from two to one or vice versa. This is not unexpected, given the conditions of positivity and realness of the solutions.

We note that this notion of generic network is different from the one we proposed in Definition 7.2. It is interesting to point out that at the image points corresponding to the resultant $K = 0$ and/or one (or both) discriminants of the numerator or denominator equal to zero (i.e., $B^2 - 4AC = 0$ and/or $E^2 - 4DF = 0$) the map ψ_Γ can become infinite-to-one, as can be seen from Table 6.11.[1] This means that for such image points there is an infinite number of combinations of network parameters which lead to the same given impedance (image point in $\mathbb{P}^5$). When the resultant K is equal to zero, there is a pole-zero cancellation in the impedance (or two if $B^2 - 4AC = E^2 - 4DF = 0$), which results in the impedance becoming bilinear (or constant). It can also be observed that in some of the subfamilies in which this occurs, the above conditions lead to a bridge balancing for some network in the subfamily. We note, however, based on our analysis of the catalogue, that this occurs only on a subset of the realizability set of dimension $d - 2$ (or less) if the realizability set has dimension d. Therefore we can say that for all networks in the catalogue (which are generic according to Definition 7.2) the map ψ_Γ is finite-to-one almost everywhere.

[1] For network #70 in equivalence class $\mathrm{V}_{\mathrm{H}}^1$ it can be easily verified that when conditions $\lambda_1 = \lambda_2 = 0$, $\rho_1 = 0$ hold, then $B^2 - 4AC = E^2 - 4DF = 0$ and $K \neq 0$. Similar considerations can be made for network #95 in equivalence class $\mathrm{V}_{\mathrm{H}}^2$.

The following claim can also be found in [59].

Claim. *The following are equivalent for a network $\Gamma \in X$ with associated graph G_Γ:*

1. *Γ is SSP.*
2. *The resultant of $a_\Gamma(s)$ and $b_\Gamma(s)$ is a product of monomials in the parameters.*
3. *Each coordinate of the inverse to ψ is expressible as a ratio of "invariants" (entries of the adjoint of the Sylvester matrix).*

The equivalence of 1 and 2 does not hold, since it can be easily calculated that the orbit of networks #69, #66, #90, #94 in subfamily $\mathrm{V_E}$ and of networks #40, #33, #24, #48 in subfamily $\mathrm{IV_C}$ both consist of networks which are SSP but whose resultants are not the product of monomials. Specifically, it can be checked that the resultant for networks #69 and #40 equals $R_1^2 L_1^3 C_1 (R_1 + R_2)^2$ in both cases. On the other hand, the equivalence of 1 and 3 does appear to hold, in light of the analysis carried out in this work, with a few remarks. In the notation of the present monograph, the biquadratic impedance

$$Z(s) = \frac{As^2 + Bs + C}{Ds^2 + Es + F}$$

has the following associated Sylvester matrix, which can be obtained from the definition of Sylvester matrix given in Section 3.3:

$$S = \begin{bmatrix} D & E & F & 0 \\ A & B & C & 0 \\ 0 & D & E & F \\ 0 & A & B & C \end{bmatrix}.$$

The adjoint (or adjugate) matrix of S is

$$\mathrm{adj}(S) = \begin{bmatrix} \lambda_4 & \lambda_1 & -C(BF-CE) & F(BF-CE) \\ A(BF-CE) & -D(BF-CE) & C(AF-CD) & -F(AF-CD) \\ -A(AF-CD) & D(AF-CD) & -C(AE-BD) & F(AE-BD) \\ A(AE-BD) & -D(AE-BD) & \lambda_2 & \lambda_3 \end{bmatrix},$$

and one can verify from Tables 6.5, 6.7, and 6.9 and from Theorems A.1–A.10 that, for all SSP networks in the catalogue,[2] the network elements can all be expressed as a ratio of entries of $\mathrm{adj}(S)$, while this is not the case for networks which are not SSP. Note from Tables 6.5 and 6.7 that, for equivalence classes $\mathrm{III_B^1}$, $\mathrm{IV_A^2}$, and $\mathrm{IV_C^2}$, we have $\lambda_2|_{D=0} = A(BE - AF)$ and the result still holds. Similar considerations hold for the other equivalence classes in these three subfamilies. Also note that the equivalence of 1 and 3 holds only if we allow multiplication by coefficients $A, B, \ldots, F$ and by the resultant $K = -\det(S)$.

7.5 ▪ A note on d-invariance of RLC networks

As mentioned in Section 5.2.2, it is the case that, within the catalogue, all networks which are d-invariant (i.e., left unchanged under duality, namely networks #3 and #108) are also invariant under the actions s and p. In other words, any network in the catalogue

[2] See Table 6.3 for a list of subfamilies which contain SSP networks.

which is 2-isomorphic to its dual is also 2-isomorphic to the networks obtained through the s and p transformations. Here we show that this property does not hold in general.

In Figure 7.5 a first example is given of an orbit of networks which are d-invariant but not s- and p-invariant. By using a Y-Δ transformation in two ways on each network, a resistor can be eliminated to produce the orbit of four nonisomorphic networks in Figure 7.6. We observe that in the latter there is a Y-Δ transformation connecting two pairs of networks, which are therefore equivalent. We now show that there is a further equivalence which allows us to conclude that all four networks in Figure 7.6 are equivalent.

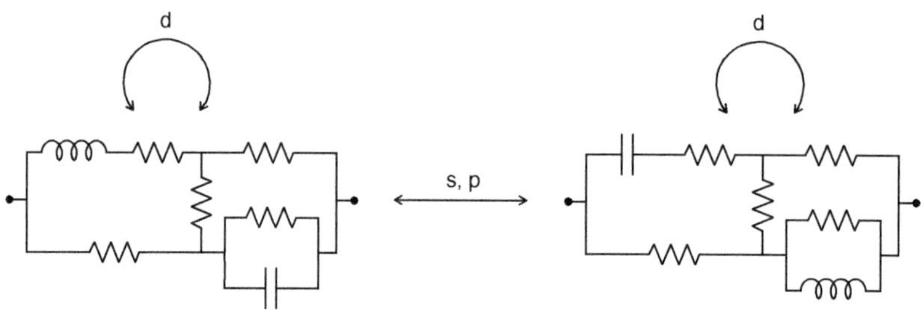

Figure 7.5. *Example of* d*-invariant networks which are not* s*- or* p*-invariant.*

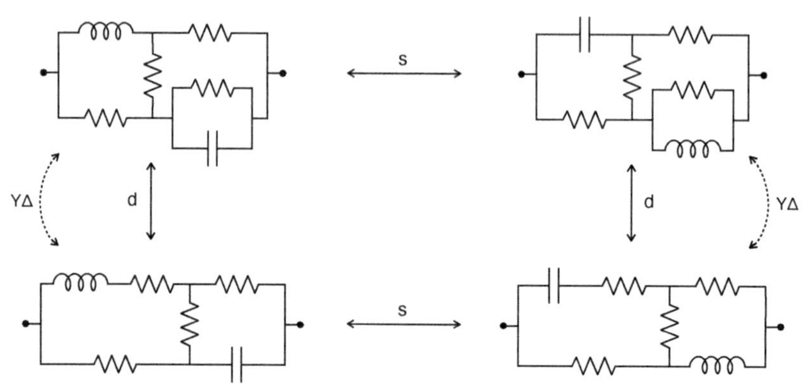

Figure 7.6. *Orbit of nonisomorphic, equivalent networks which realize a biquadratic impedance.*

It can be shown that the networks in Figure 7.7 are related by the following transformation:

$$a' = \frac{c(ac+P)}{d(b+c)}, \quad b' = \frac{Pb\,(ac+P)}{d(b+c)\big[a(b+c)+P\big]}, \quad c' = \frac{aP}{a(b+c)+P},$$

$$d' = \frac{(ac+P)^2}{(a+b)\big[a(b+c)+P\big]}, \quad e' = \frac{f\,P^2}{d^2(b+c)^2}, \quad f' = \frac{e(ac+P)^2}{\big[a(b+c)+P\big]^2},$$

where $P = bc + bd + cd$

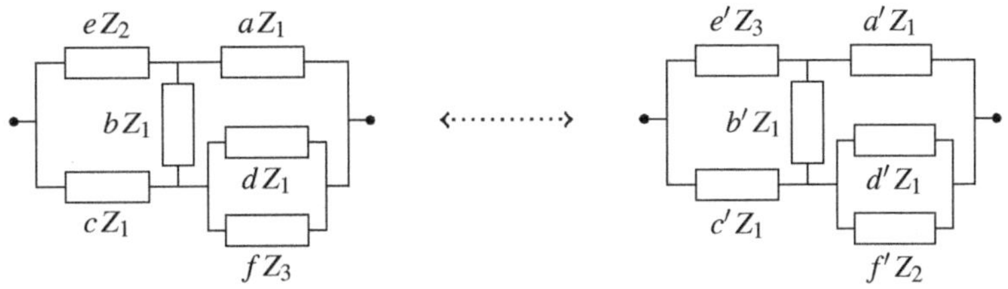

Figure 7.7. *New equivalence between RLC networks.*

for any real positive numbers a, b, ..., f, where $Z_1(s)$, $Z_2(s)$, and $Z_3(s)$ are arbitrary impedances. The inverse transformation is given by the same expressions above, by replacing a, b, ..., f with a', b', ..., f'.

Using this new equivalence on the networks of Figure 7.6 we can conclude that the four networks, although not 2-isomorphic, are all equivalent. This suggests that a weaker property might hold in general, namely that a network being equivalent to its dual always implies that it is also equivalent to the networks obtained through frequency inversion and the p transformation. It was not possible to find a counterexample to this among the networks with up to six elements. The higher-order networks shown in Figure 7.8, however, turned out to be an example of networks which are d-invariant but not s- or p-invariant and *not* equivalent, as we will now show, thus refuting this last conjecture.

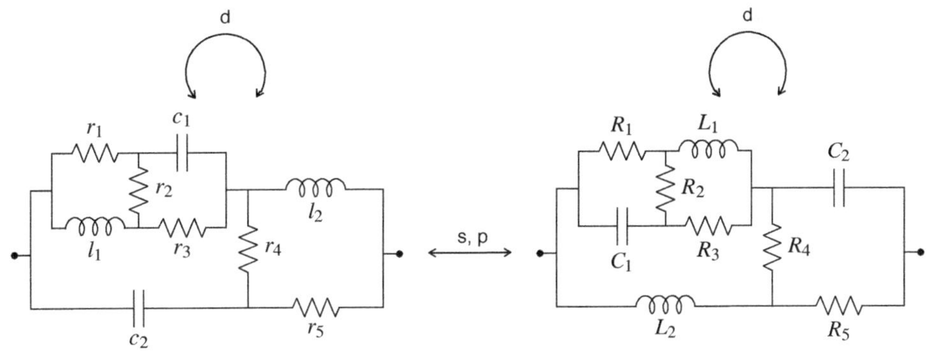

Figure 7.8. *Example of* d*-invariant networks in the same orbit which are not equivalent.*

In order to prove that the networks in Figure 7.8 are not equivalent it is sufficient to find an impedance in the realizability set of one network which is not realizable by the other network. This is the case for the impedance obtained by setting $r_1 = r_4 = 2$ and $r_2 = r_3 = r_5 = l_1 = l_2 = c_1 = c_2 = 1$ in the first network, which gives the biquartic impedance

$$Z_1(s) = \frac{n_1(s)}{d_1(s)} = \frac{8s^4 + 36s^3 + 57s^2 + 36s + 9}{8s^4 + 36s^3 + 51s^2 + 36s + 9}, \tag{7.3}$$

where $n_1(s)$ and $d_1(s)$ have no common factor. An expression $Z_2(s) = n_2(s)/d_2(s)$ can be obtained for the impedance of the second network, where $n_2(s)$ and $d_2(s)$ are fourth-order polynomials in s whose coefficients depend on the network parameters R_1,

R_2, L_1, etc. By setting $Z_1(s) = Z_2(s)$ we obtain the eighth-order polynomial equation

$$n_1(s)d_2(s) - n_2(s)d_1(s) = 0\,,$$

which is satisfied if the coefficients of each power of s are equal to zero. This leads to nine polynomial equations in the nine network parameters R_1, R_2, L_1, etc. Through an elimination procedure we obtain $R_3 = R_5 = 1$, $L_2 = C_2$, $C_1 = L_1$ and the following expressions for R_1, C_2, R_4 which are rational functions of L_1 and R_2 only:

$$\begin{aligned}
R_1 &= -\frac{3L_1^2R_2 + 4R_2 + 4}{3L_1^2 - 4R_2 - 4}\,,\\
C_2 &= \frac{-8L_1(R_2+1)}{3L_1^2(R_2-1) - 18L_1(R_2+1) + 8(R_2+1)}\,,\\
R_4 &= \frac{3C_2^2(1-R_1) + 2R_1}{3C_2^2(R_1-1) - 2}\,,
\end{aligned}$$

where L_1 and R_2 still have to satisfy the pair of polynomial equations

$$\begin{aligned}
&27(R_2-1)^2\,L_1^4 - 108(R_2^2-1)\,L_1^3 + 12(R_2+1)(19R_2-1)\,L_1^2\\
&- 288(R_2+1)^2\,L_1 + 128(R_2+1)^2 = 0\,,
\end{aligned} \tag{7.4}$$

$$\begin{aligned}
&-9(R_2-1)^3\,L_1^5 + 108(R_2+1)(R_2-1)^2\,L_1^4 - 12(R_2^2-1)(31R_2+23)\,L_1^3\\
&+ 48(17R_2-1)(R_2+1)^2\,L_1^2 - 32(29R_2+25)(R_2+1)^2\,L_1\\
&+ 384(R_2+1)^3 = 0\,.
\end{aligned} \tag{7.5}$$

For a given value of R_2, the two equations will have a common root if and only if their resultant (which is a function of R_2 only) is zero. The resultant of (7.4) and (7.5) is a constant multiple of

$$(1861R_2^3 + 13371R_2^2 + 32475R_2 + 17093)(R_2-1)^6(R_2+1)^{13}\,,$$

and it is easily seen that its only real, positive solution is $R_2 = 1$. For $R_2 = 1$, (7.4) and (7.5) reduce to the quadratics

$$\begin{aligned}
432\,L_1^2 - 1152\,L_1 + 512 = 0\,,\\
3072\,L_1^2 - 6912\,L_1 + 3072 = 0\,,
\end{aligned}$$

which have no common root. We can therefore conclude that no real, positive values exist for R_1, R_2, L_1, etc., which make $Z_2(s)$ equal the candidate impedance (7.3), and hence the two networks of Figure 7.8 are not equivalent.

7.6 ▪ Six-element networks with four resistors

In light of the discussion in Section 7.5 and the introduction of the new equivalence shown in Figure 7.7, we now consider the entire class of networks containing two reactive elements and four resistors. Although it was shown that additional resistors beyond three do not expand the class of functions that are realized by the Ladenheim catalogue (see Section 3.2 and [67, 48]), it is still interesting to explore the structure of this class and possibly uncover further equivalences.

Within the Ladenheim catalogue there are 25 basic graph structures with five edges, of which 24 are SP graphs and only one is a bridge graph (graph V in Appendix B). We consider now all the distinct graph structures with six edges. There are 72 such graphs, of which 66 are SP [70] and six are bridge graphs. Half of the 66 SP graphs are presented in [88, 89] (the other half being obtained through duality), while the six bridge graphs are shown in Figure 7.9 and can be obtained from [26].

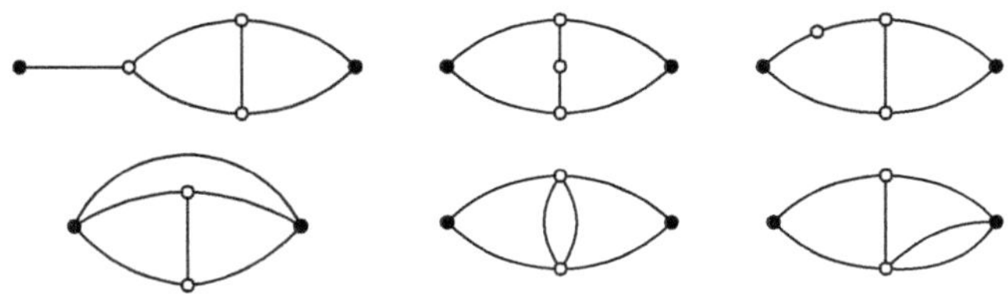

Figure 7.9. *Bridge graphs with six edges. The bottom three graphs are the duals of the top three graphs.*

Considering all the essentially distinct RLC networks which can be obtained by populating the edges with four resistors and two reactive elements, and eliminating the networks which further reduce through a Zobel or a Y-Δ transformation, leads to a class of 52 networks, of which 44 are bridge networks and eight are SP networks. These 52 networks can be analyzed and grouped according to the classification tools presented in Chapters 5 and 6. Specifically, we can partition the set into 15 orbits and, identifying all the equivalences that result from Zobel and Y-Δ transformations, 28 disjoint sets of equivalent networks. Considering the finest common coarsening of these two partitions leads to nine subfamilies. At this point it is not clear whether new equivalences might lead some of these subfamilies to coalesce. From Section 7.5 we already know of one further equivalence. We now investigate whether there are any additional equivalences, by comparing the realizability regions of such networks.

The realizability regions were computed numerically and plotted for one network in each subfamily. We note that letting one of the four resistances go to zero or infinity leads to networks of the Ladenheim catalogue. It was possible to verify that for all the networks with one inductor and one capacitor the realizability region is obtained as the union of the realizability regions of the five-element networks to which the network can reduce. An example is given in Figure 7.10.

Looking at the realizability region of the networks in the orbit shown in Figure 7.11, and noting that it is symmetric with respect to the $U = V$ bisector, suggests that these networks are all equivalent to their p-transform. In fact, it can be shown that the networks in Figure 7.12 are equivalent. In particular, they are related by the transformation

$$a' = \frac{T(a+e+f)}{a(c+f)}, \quad b' = \frac{(a+e+f)^2 b}{(e+f)^2}, \quad c' = \frac{(a+e+f)\big[T+e(a+c)\big]}{(e+f)^2},$$

$$d' = \frac{d(a+e+f)^2\,\big[T+e(a+c)\big]^2}{(T+ce)^2\,(e+f)^2}, \quad e' = \frac{(a+e+f)^2\,Te}{f(e+f)(T+ce)},$$

$$f' = \frac{T(a+e+f)\big[T+e(a+c)\big]}{c(e+f)(T+ce)}, \quad \text{where } T = ac+af+cf,$$

$$\left[a = \frac{a'c'f'(e'+f')}{(a'+e'+f')\,T'}, \quad b = \frac{(a')^2\, b'(c'+f')^2}{(T')^2}, \quad c = \frac{c'(c'+f')(T'+c'e')(a')^2}{(T')^2\,(a'+e'+f')}, \right.$$

$$d = \frac{(a')^2\, d'(c'+f')^2(T'+c'e')^2}{(T')^2\left[T'+e'(c'+f')\right]^2}, \quad e = \frac{(a')^2\, e'(c'+f')^2}{T'\left[T'+e'(c'+f')\right]},$$

$$\left. f = \frac{(a')^2\, f'(c'+f')(T'+c'e')^2}{(a'+e'+f')T'\left[T'+e'(c'+f')\right]}, \quad \text{where } T' = a'c' + a'f' + c'f' \right]$$

for any real positive numbers a, b, etc., where $Z_1(s)$, $Z_2(s)$ and $Z_3(s)$ are arbitrary impedances.

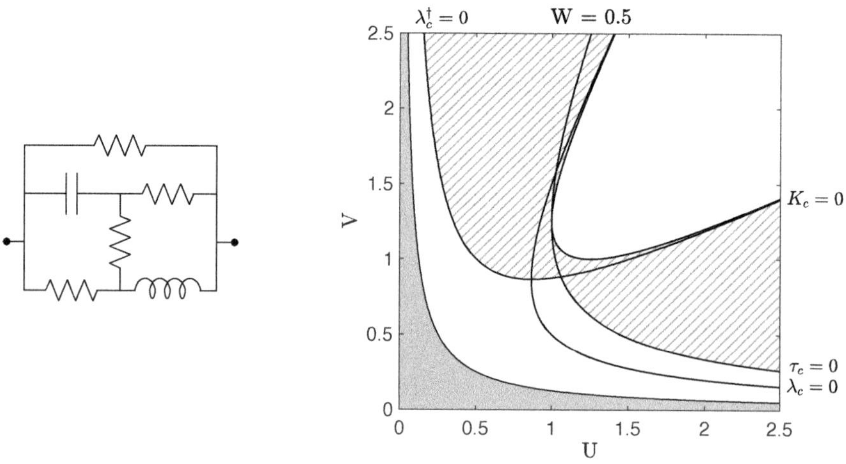

Figure 7.10. *Two-reactive, six-element network and corresponding realizability region for $W = 0.5$. Since the network is* s*-invariant, the region is the same for $W = 2$. Allowing one of the four resistances to be zero or infinity leads to networks of equivalence classes V_I, V_E^2, V_E^3, V_F^1, and V_G^2. It can be verified from the plots in Section* 6.6 *that the realizability region shown above is the union of the realizability regions of the above-mentioned five-element equivalence classes.*

Therefore two new equivalences among RLC networks surfaced in this analysis: the equivalence shown in Figure 7.7 and the one in Figure 7.12. These led to the new equivalences shown in Figures 7.6 and 7.11 in the class of two-reactive, four-resistor networks. Using numerical analysis it was shown that there are no further equivalences in this class. Furthermore the two new equivalences did not cause any of the initial nine subfamilies to coalesce. Hence we can state the following proposition.

Proposition 7.7. *The* 52 *networks in the class of two-reactive, four-resistor networks can be partitioned into* 23 *equivalence classes, which form nine subfamilies. (Table* 7.1 *provides more detail on the structure of the subfamilies.)*

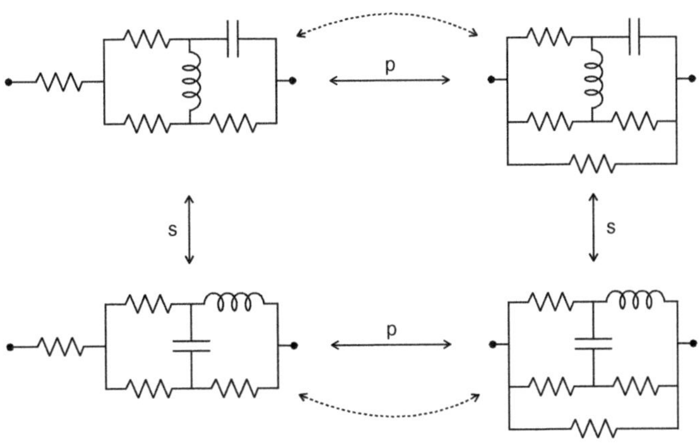

Figure 7.11. *Orbit of six-element networks related by a new equivalence.*

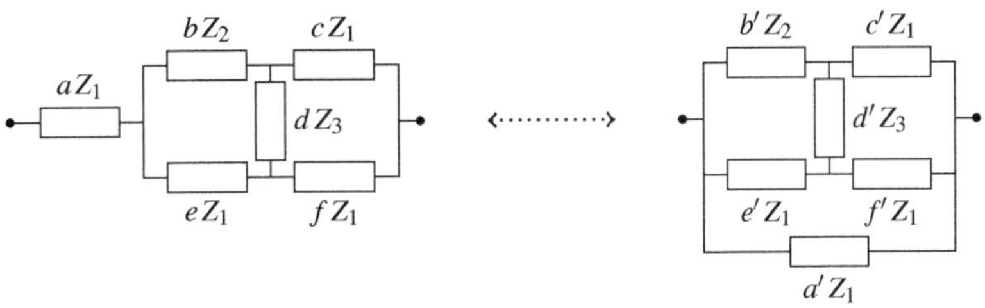

Figure 7.12. *New equivalence between RLC networks.*

Table 7.1. *Composition of each of the nine subfamilies within the class of two-reactive, six-element networks.*

		# Eq. classes	# Orbits	# Networks
Subfamilies of RL/RC networks	subf. 1	2	2	8
	subf. 2	4	1	4
	subf. 3	4	1	4
	subf. 4	2	1	4
Subfamilies of RLC networks	subf. 5	2	5	14
	subf. 6	2	2	8
	subf. 7	2	1	2
	subf. 8	4	1	4
	subf. 9	1	1	4
	Total	23	15	52

Chapter 8
Conclusions

This monograph is a continuation of recent research in the field of passive network synthesis. A substantial part of the monograph consists of a detailed analysis of the structure of the Ladenheim catalogue. In addition, a summary is provided of relevant classical and recent results, and possible applications in the mechanical domain are surveyed. The new material is a continuation of recent work which is motivated by the need for economical realizations in the mechanical domain. This program of work can also be seen in the context of the challenge of Foster to seek better methods of RLC synthesis (see Chapter 1).

The main new results of the monograph relate to the Ladenheim catalogue, which is the class of networks consisting of at most two reactive elements and at most three resistors. Our goal has been to obtain a complete understanding of the enumerative approach to network synthesis in this simplest nontrivial case. The task entailed an analysis of the realization power of every circuit in the catalogue. The results are summarized in diagrammatic form in (i) Figures 6.1–6.3 (orbits, equivalence classes, and subfamilies), (ii) Table 6.11 (realizability conditions for all equivalence classes), and (iii) Figures 6.4–6.14 (realizability regions for five-element networks). To aid the classification several new ideas were introduced which we now summarize.

A previous notion of "network quartet" was replaced by the orbits induced by the group actions s (frequency inversion) and p (circuit dual without element dual), together with d (dual) and e (identity). The use of s and p as the primary representatives is new in the present work and was suggested by the more convenient grouping of the equivalence classes within the subfamilies.

A formal notion of "realizability set" was introduced as a semi-algebraic set in the space of impedance parameters in order to compare the realization power of all the networks (Section 5.2.1). The key findings of this monograph stem from the realization theorems that were derived in Appendix A. For all circuits of the catalogue it was possible to derive necessary and sufficient conditions which allowed the corresponding semi-algebraic sets to be compared. A main result of the monograph is that there are no new equivalences within the catalogue that are not induced by the classical Zobel or Y-Δ transformations. An analysis of the class of six-element networks with four resistors, however, led to two new equivalences being found. Furthermore, it was shown that the classical Cauer and Foster forms are almost but not completely equivalent and may be termed quasi-equivalent. The determination of the realizability sets for all networks in the catalogue also allowed the question of the smallest generating set for the catalogue to be settled.

A new notion of "generic network" was introduced based on the notion of dimension of a semi-algebraic set (Definition 7.2). It was shown that the 108 networks of the catalogue comprise the complete set of "essentially distinct," generic five-element networks with at most two reactive elements.

Finally, much structure has surfaced from the analysis of this apparently simple class of five-element networks. Whether further patterns could be revealed is an open question. Nevertheless the structure that has been revealed will no doubt inform a more general enumerative approach to RLC synthesis that is a topic of future research.

Appendix A
Realization theorems

A.1 ▪ Equivalence class IV_B^1

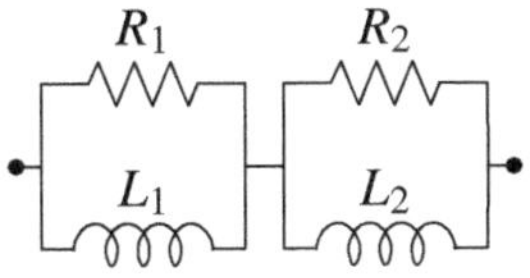

Figure A.1. *Network #38 from subfamily IV_B. By Lemma 8 in [47] it can only realize regular impedances.*

Theorem A.1. *Let A, B, C, D, E, $F \geq 0$. The positive-real biquadratic impedance* (3.1) *can be realized as in Figure* A.1, *with R_1, R_2, L_1, and L_2 positive and finite, if and only if*

$$A, B, D, E, F > 0\,, \tag{A.1}$$

$$C = 0\,, \tag{A.2}$$

and either

$$K < 0 \tag{A.3}$$

or

$$K = 0 \text{ and } E^2 - 4DF = 0\,, \tag{A.4}$$

where K is defined in (3.3). *If condition* (A.3) *is satisfied, then R_1 and R_2 are the two roots of the quadratic equation in x*

$$-D(E^2 - 4DF)\,x^2 + A(E^2 - 4DF)\,x + (A^2F - ABE + B^2D) = 0\,, \tag{A.5}$$

while if (A.4) *holds, then R_1 and R_2 are any two positive values such that $R_1 + R_2 = A/D$. The two inductances can be obtained as*

$$L_1 = \frac{R_1(B - ER_2)}{F(R_1 - R_2)}\,, \tag{A.6}$$

$$L_2 = \frac{R_2(B - ER_1)}{F(R_2 - R_1)} \tag{A.7}$$

if $R_1 \neq R_2$ *and as the two roots of the quadratic equation in* y

$$2EF\,y^2 - 2BE\,y + AB = 0 \tag{A.8}$$

if $R_1 = R_2$. *Due to the symmetry of the network, the two solutions of the quadratic do not lead to two properly distinct solutions.*

Proof. *Necessity.* The impedance of the network shown in Figure A.1 is a biquadratic, which can be computed as

$$Z(s) = \frac{n(s)}{d(s)}\,, \tag{A.9}$$

where

$$\begin{aligned} n(s) &= L_1L_2(R_1+R_2)s^2 + R_1R_2(L_1+L_2)s\,,\\ d(s) &= L_1L_2s^2 + (L_1R_2+L_2R_1)s + R_1R_2\,. \end{aligned}$$

Equating impedance (A.9) with (3.1) we obtain, for a positive constant k,

$$L_1L_2(R_1+R_2) = kA\,, \tag{A.10}$$

$$R_1R_2(L_1+L_2) = kB\,, \tag{A.11}$$

$$0 = kC\,, \tag{A.12}$$

$$L_1L_2 = kD\,, \tag{A.13}$$

$$L_1R_2 + L_2R_1 = kE\,, \tag{A.14}$$

$$R_1R_2 = kF\,, \tag{A.15}$$

which are a set of necessary and sufficient conditions for (3.1) to be realized as in Figure A.1. It can be calculated that

$$K = -k^{-4}\,L_1L_2R_1^2R_2^2(L_1R_2 - L_2R_1)^2 \leq 0\,, \tag{A.16}$$

$$E^2 - 4DF = k^{-2}\,(L_1R_2 - L_2R_1)^2 \geq 0\,, \tag{A.17}$$

from which we can conclude that

$$K = 0 \;\Leftrightarrow\; E^2 - 4DF = 0 \;\Leftrightarrow\; L_1R_2 = L_2R_1\,, \tag{A.18}$$

and hence (A.1)–(A.4) are necessary.

Sufficiency. We now assume that (A.1)–(A.2) hold and either (A.3) or (A.4). We show that we can find R_1, R_2, L_1, L_2 positive which satisfy (A.10)–(A.15) with $k > 0$.

Eliminating R_1R_2 from (A.11) and (A.15) we obtain

$$L_2 = B/F - L_1\,, \tag{A.19}$$

and from (A.14) and (A.19) we then obtain

$$k = \frac{L_1R_2 + L_2R_1}{E} = \frac{FL_1(R_2 - R_1) + BR_1}{EF}\,. \tag{A.20}$$

Substituting (A.20) into (A.15) and solving for L_1 we obtain (A.6), and from (A.19) and (A.6) we obtain (A.7), providing $R_1 \neq R_2$ (which we assume for the time being). Eliminating L_1L_2 from (A.10) and (A.13) we obtain

$$R_2 = A/D - R_1\,. \tag{A.21}$$

Substituting the values thus obtained for L_1, L_2, R_2, and k into (A.13) we get the quadratic (A.5) in R_1, and we note from the expression in Table 6.10 that

$$K|_{C=0} = F(A^2F - ABE + B^2D)\,, \tag{A.22}$$

which is the third coefficient in (A.5). It is easily seen that the sum of the two roots of (A.5) is A/D, and hence R_2 can be obtained as the other root of the same quadratic, as stated in the theorem.

It can be easily verified that

$$A^2(E^2 - 4DF) = (AE - 2BD)^2 - 4DK/F\,, \tag{A.23}$$

and hence condition (A.3) implies $E^2 - 4DF > 0$. We can therefore conclude that if (A.3) holds, the first and third coefficients in (A.5) are strictly negative, while the second coefficient is strictly positive. It also follows from $E^2 - 4DF > 0$ that the discriminant of (A.5),

$$\Delta_x = (E^2 - 4DF)(AE - 2BD)^2\,,$$

is greater than or equal to zero, and hence the quadratic has two real positive roots. Since we are assuming $R_1 \neq R_2$, the discriminant will necessarily be nonzero, and hence $AE - 2BD \neq 0$ in this case. We finally verify that the values obtained from (A.6) and (A.7) for L_1 and L_2 are always positive. Since R_1 and R_2 are the two distinct roots of the quadratic (A.5), L_1 and L_2 will be positive if B/E lies strictly between the two roots of (A.5) (from the expressions in (A.6) and (A.7)). This is true, providing

$$-D(E^2 - 4DF)\,x^2 + A(E^2 - 4DF)\,x + (A^2F - ABE + B^2D) > 0 \tag{A.24}$$

for $x = B/E$. After some manipulation, inequality (A.24) can be reduced to $F(AE - 2BD)^2/E > 0$, which holds in this case.

If $R_1 \neq R_2$ and conditions (A.4) hold, then all the coefficients in (A.5) are zero and any value of x solves the quadratic. Therefore R_1 can be chosen arbitrarily, providing $R_2 = A/D - R_1 > 0$. From (A.23), conditions (A.4) imply $AE - 2BD = 0$. It was observed above that L_1 and L_2 are positive if B/E lies strictly between the two resistances. Without loss of generality we can assume $R_1 < R_2$ and, since $R_1 + R_2 = A/D$, it follows that $R_1 < A/(2D) < R_2$. Since $A/(2D) = B/E$, it then follows that $R_1 < B/E < R_2$, and hence L_1 and L_2 are positive. We have therefore shown that (A.10)–(A.15) can be satisfied for $R_1, R_2, L_1, L_2 > 0$ (under the assumption that $R_1 \neq R_2$ and either (A.3) or (A.4) holds).

We now turn to the case that $R_1 = R_2$. Equations (A.19) and (A.20) still hold, and eliminating L_1L_2 from (A.10) and (A.13) we obtain $R_1 = R_2 = A/(2D)$. Equation (A.15), for $R_1 = R_2$, results in the identity

$$AE - 2BD = 0\,. \tag{A.25}$$

Substituting the values thus obtained for R_1, R_2, L_2, and k into (A.13) we obtain the quadratic (A.8) in L_1. It can be easily seen that the sum of the two roots of (A.8) is B/F, and hence L_2 can also be obtained as a root of (A.8).[3] It can also be easily verified that

$$A(2AF - BE) = 2K/F + B(AE - 2BD)\,,$$

[3] Again we note that, since $R_1 = R_2$, swapping the order of the inductors does not lead to two properly distinct solutions.

which, by virtue of (A.25), reduces to $A(2AF - BE) = 2K/F$. The discriminant of (A.8),

$$\Delta_y = -4BE(2AF - BE) = -8BEK/(AF),$$

is therefore greater than or equal to zero if either (A.3) or (A.4) holds. Given the signs of the coefficients in (A.8), we can therefore conclude that the two roots of the quadratic are positive and real. We finally note that replacing $x = A/(2D)$ in the quadratic (A.5) we obtain $(AE - 2BD)^2/(4D)$, which by virtue of (A.25) is zero. Hence, even in this case R_1 and R_2 can be obtained as the two roots of (A.5). □

A.2 ▪ Equivalence class V_A^1

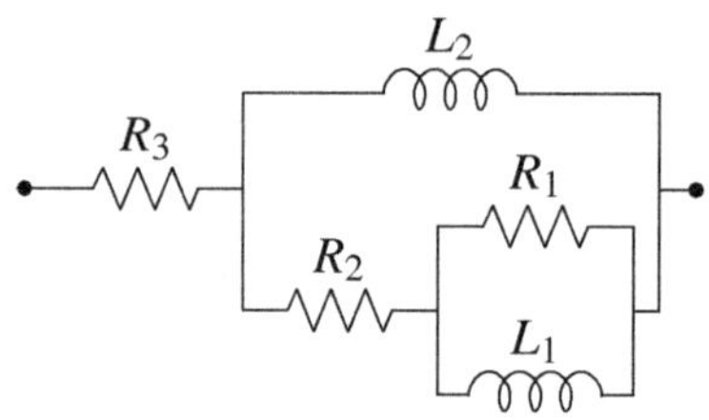

Figure A.2. *Network #52 from subfamily V_A. By Theorem* 1 *in* [47] *it can only realize regular impedances.*

Theorem A.2. *The positive-real biquadratic impedance* (3.1) *with A, B, C, D, E, $F > 0$ can be realized as in Figure* A.2*, with R_1, R_2, R_3, L_1, and L_2 positive and finite, if and only if*

$$AF - CD > 0, \tag{A.26}$$

$$K < 0. \tag{A.27}$$

If conditions (A.26) *and* (A.27) *are satisfied, then*

$$R_1 = \frac{-K}{D\lambda_1}, \tag{A.28}$$

$$R_2 = \frac{(BF - CE)^2}{F\lambda_1}, \tag{A.29}$$

$$R_3 = \frac{C}{F}, \tag{A.30}$$

$$L_1 = \frac{-K(BF - CE)}{\lambda_1^2}, \tag{A.31}$$

$$L_2 = \frac{BF - CE}{F^2}, \tag{A.32}$$

where λ_1 and K are defined in Table 6.10.

Proof. Note this result is given without proof in [47, Appendix A], with R_1 and R_3 interchanged. We provide here a proof for convenience.

Necessity. The impedance of the network shown in Figure A.2 is a biquadratic, which can be computed as

$$Z(s) = \frac{n(s)}{d(s)}, \tag{A.33}$$

where

$$n(s) = L_1L_2(R_1 + R_2 + R_3)s^2 + \big(L_1R_3(R_1 + R_2) + L_2R_1(R_2 + R_3)\big)s + R_1R_2R_3,$$
$$d(s) = L_1L_2s^2 + \big(L_2R_1 + L_1(R_1 + R_2)\big)s + R_1R_2.$$

Equating impedance (A.33) with (3.1), we obtain, for a positive constant k,

$$L_1L_2(R_1 + R_2 + R_3) = kA, \tag{A.34}$$
$$L_1R_3(R_1 + R_2) + L_2R_1(R_2 + R_3) = kB, \tag{A.35}$$
$$R_1R_2R_3 = kC, \tag{A.36}$$
$$L_1L_2 = kD, \tag{A.37}$$
$$L_2R_1 + L_1(R_1 + R_2) = kE, \tag{A.38}$$
$$R_1R_2 = kF, \tag{A.39}$$

which are a set of necessary and sufficient conditions for (3.1) to be realized as in Figure A.2. It can be calculated that

$$AF - CD = k^{-2}\, R_1R_2L_1L_2\,(R_1 + R_2) > 0, \tag{A.40}$$
$$K = -k^{-4}\, R_1^4R_2^2L_1L_2^3 < 0, \tag{A.41}$$

and hence (A.26) and (A.27) are necessary.

Sufficiency. Given a positive-real impedance (3.1) with $A, B, C, D, E, F > 0$ satisfying conditions (A.26) and (A.27) we now show that we can find R_1, R_2, R_3, L_1, L_2 positive which satisfy (A.34)–(A.39) with $k > 0$.

From (A.39) we obtain $k = R_1R_2/F$ and eliminating the term R_1R_2 from (A.36) and (A.39) we get (A.30). Eliminating L_1 from (A.35) and (A.38) we obtain (A.32), while eliminating the term L_1L_2 from (A.34) and (A.37) and solving for R_2 we get

$$R_2 = \frac{AF - CD - DFR_1}{DF}. \tag{A.42}$$

Solving for L_1 from (A.37) using expressions obtained we find

$$L_1 = \frac{R_1\,(AF - CD - DFR_1)}{BF - CE}. \tag{A.43}$$

We now have expressions for R_2, R_3, L_1, L_2, and k which only contain R_1 together with A, B, C, D, E, F. Substituting such expressions into (A.38) and solving for R_1 we obtain (A.28) and, substituting the latter into (A.42) and (A.43), we get (A.29) and (A.31), respectively.

Conditions (A.26) and (A.27) imply $BF - CE > 0$, from which it can be shown, using again (A.26) and (A.27), that $\lambda_1 > 0$ (note this also follows from [47, Lemma 7]). Hence all network elements are positive. □

A.3 ▪ Equivalence class V_B^1

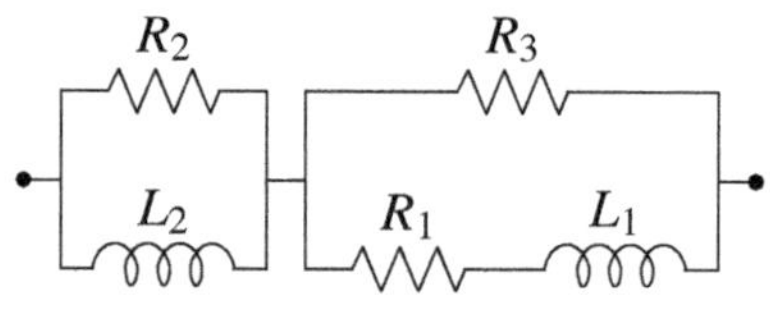

Figure A.3. *Network #59 from subfamily V_B. By Theorem* 1 *in* [47] *it can only realize regular impedances.*

Theorem A.3. *The positive-real biquadratic impedance* (3.1) *with A, B, C, D, E, $F > 0$ can be realized as in Figure* A.3, *with R_1, R_2, R_3, L_1, and L_2 positive and finite, if and only if*

$$AF - CD > 0 \tag{A.44}$$

and either

$$K < 0 \tag{A.45}$$

or

$$K = 0 \text{ and } E^2 - 4DF = 0\,, \tag{A.46}$$

where K is defined in (3.3). *In case* (A.45), *R_2 is either of the two positive roots of the quadratic equation in x*

$$c_1\,x^2 + c_2\,x + c_3 = 0\,, \tag{A.47}$$

where

$$\begin{aligned} c_1 &= -DF(E^2 - 4DF)\,, \\ c_2 &= (E^2 - 4DF)(AF - CD)\,, \\ c_3 &= K\,, \end{aligned}$$

while, in case (A.46), *R_2 may take any positive value that is strictly less than $(AF - CD)/(DF)$. The other network elements are given by*

$$R_3 = A/D - R_2\,, \tag{A.48}$$

$$R_1 = \frac{CR_3}{FR_3 - C}\,, \tag{A.49}$$

$$L_1 = \frac{D(R_1R_2 + R_2R_3 - R_3^2)}{B - ER_3}\,, \tag{A.50}$$

$$L_2 = \frac{DR_2(R_1 + R_3)}{FL_1}\,, \tag{A.51}$$

unless $R_2 = (AE - BD)/(DE)$, in which case L_1 is either of the two positive roots of the quadratic equation in y

$$d_1\,y^2 + d_2\,y + d_3 = 0\,, \tag{A.52}$$

where

$$\begin{aligned} d_1 &= E^2F(BF - CE)^2\,, \\ d_2 &= -B^2E^2F(BF - CE)\,, \\ d_3 &= B^4DF\,, \end{aligned}$$

and the other elements are still given by (A.48)–(A.49) *and* (A.51).

Proof. *Necessity.* The impedance of the network shown in Figure A.3 is a biquadratic, which can be computed as

$$Z(s) = \frac{n(s)}{d(s)}, \tag{A.53}$$

where

$$n(s) = L_1 L_2 (R_2 + R_3) s^2 + \big(L_1 R_2 R_3 + L_2 (R_1 R_2 + R_1 R_3 + R_2 R_3)\big) s + R_1 R_2 R_3,$$
$$d(s) = L_1 L_2 s^2 + \big(L_1 R_2 + L_2 (R_1 + R_3)\big) s + R_2 (R_1 + R_3).$$

Equating impedance (A.53) with (3.1), we obtain, for a positive constant k,

$$L_1 L_2 (R_2 + R_3) = kA, \tag{A.54}$$
$$L_1 R_2 R_3 + L_2 (R_1 R_2 + R_1 R_3 + R_2 R_3) = kB, \tag{A.55}$$
$$R_1 R_2 R_3 = kC, \tag{A.56}$$
$$L_1 L_2 = kD, \tag{A.57}$$
$$L_1 R_2 + L_2 (R_1 + R_3) = kE, \tag{A.58}$$
$$R_2 (R_1 + R_3) = kF, \tag{A.59}$$

which are a set of necessary and sufficient conditions for (3.1) to be realized as in Figure A.3. It can be calculated that

$$AF - CD = k^{-2}\, L_1 L_2 R_2 (R_1 R_2 + R_2 R_3 + R_3^2) > 0, \tag{A.60}$$
$$K = -k^{-4}\, L_1 L_2 R_2^2 R_3^2 (L_1 R_2 - L_2 (R_1 + R_3))^2 \le 0, \tag{A.61}$$
$$E^2 - 4DF = k^{-2}\, (L_1 R_2 - L_2 (R_1 + R_3))^2 \ge 0, \tag{A.62}$$

and from (A.61) and (A.62) we can conclude that

$$K = 0 \;\Leftrightarrow\; E^2 - 4DF = 0 \;\Leftrightarrow\; L_1 R_2 = L_2 (R_1 + R_3),$$

and hence (A.44)–(A.46) are necessary.

Sufficiency. We now assume that (A.44) and either (A.45) or (A.46) holds. We show that we can find R_1, R_2, R_3, L_1, L_2 positive which satisfy (A.54)–(A.59) with $k > 0$. From (A.44) and $K \le 0$ it follows that $AE - BD > 0$ and $BF - CE > 0$, which we will assume for the rest of the proof.

From (A.59) we obtain

$$k = R_2 (R_1 + R_3)/F, \tag{A.63}$$

and from (A.57) and (A.63) we get (A.51). Substituting (A.63) into (A.56) we obtain (A.49) (and it will follow below that $R_3 \neq C/F$, using (A.67) and (A.69)). Eliminating $L_1 L_2$ from (A.54) and (A.57) we get (A.48). Using (A.63) and (A.51), equations (A.55) and (A.58) reduce to

$$F R_2 R_3\, L_1^2 - B R_2 (R_1 + R_3)\, L_1 + D R_2 (R_1 + R_3)(R_1 R_2 + R_1 R_3 + R_2 R_3) = 0,$$
$$F R_2\, L_1^2 - E R_2 (R_1 + R_3)\, L_1 + D R_2 (R_1 + R_3)^2 = 0.$$

Eliminating the L_1^2 term from the first equation we obtain

$$L_1 (B - E R_3) - D (R_1 R_2 + R_2 R_3 - R_3^2) = 0. \tag{A.64}$$

Assuming $R_3 \neq B/E$ and solving for L_1 we obtain (A.50). We now have expressions for R_1, R_3, L_1, L_2, and k which only contain R_2, together with A, B, C, D, E, F. Substituting such expressions into (A.58) we get the quadratic equation (A.47) in R_2. We note that $R_3 \neq B/E$ is equivalent to $R_2 \neq (AE - BD)/(DE)$ from (A.48).

We now assume that conditions (A.44)–(A.45) hold with $R_2 \neq (AE-BD)/(DE)$. The discriminant of (A.47) is given by

$$\Delta_c = (E^2 - 4DF)\,\theta_2^2\,,$$

where $\theta_2 = AEF - 2BDF + CDE$, and it can be verified that the following identity always holds:

$$(BF - CE)(AE - BD)(E^2 - 4DF) = \theta_2^2 - E^2\,K\,. \tag{A.65}$$

From (A.65) we can conclude that if (A.44)–(A.45) hold, then $E^2 - 4DF > 0$ necessarily, and hence the first and third coefficients in (A.47) are negative while the second is positive. If $\theta_2 = 0$, then the discriminant of (A.47) is zero, and the two coincident roots of (A.47) are $R_2 = (AF - CD)/(2DF)$. The following identity always holds:

$$\frac{AE - BD}{DE} - \frac{AF - CD}{2DF} = \frac{\theta_2}{2DEF}\,, \tag{A.66}$$

from which it follows, if $\theta_2 = 0$, that

$$R_2 = \frac{AF - CD}{2DF} = \frac{AE - BD}{DE}\,,$$

which is a contradiction in this case. We can therefore conclude that $\theta_2 \neq 0$, i.e., the discriminant of (A.47) is strictly greater than zero. Therefore (A.47) has two distinct positive roots, which we denote as R_{2_A} and R_{2_B}, with $R_{2_A} < R_{2_B}$. From (A.48), R_3 is positive for both sets of solutions if $R_{2_B} < A/D$. From (A.49), making use of (A.48), and from (A.50), making use of the identities (A.48) and (A.49), we obtain

$$R_1 = \frac{CDR_3}{-DF\,R_2 + AF - CD}\,, \tag{A.67}$$

$$L_1 = \frac{DR_1R_3(-2D\,FR_2 + AF - CD)}{C(-DE\,R_2 + AE - BD)}\,. \tag{A.68}$$

We now show that

$$\begin{aligned} &R_{2_A} < \min\left\{\frac{AE - BD}{DE}, \frac{AF - CD}{2DF}\right\}, \\ &\max\left\{\frac{AE - BD}{DE}, \frac{AF - CD}{2DF}\right\} < R_{2_B} < \frac{AF - CD}{DF} < \frac{A}{D}, \end{aligned} \tag{A.69}$$

from which it will follow from (A.48), (A.67), and (A.68) that $R_3 > 0$, $R_1 > 0$, and $L_1 > 0$ for both solutions (we note, incidentally, from (A.66) that

$$\frac{AE - BD}{DE} \gtrless \frac{AF - CD}{2DF}\,,$$

depending on the sign of θ_2). We first show that $(AE - BD)/(DE)$ and $(AF - CD)/(2DF)$ always lie between the two roots of (A.47). Since (A.47) represents a parabola that opens down, the latter is true if and only if

$$c_1 \left(\frac{AE - BD}{DE}\right)^2 + c_2 \left(\frac{AE - BD}{DE}\right) + c_3 > 0\,, \tag{A.70}$$

$$c_1 \left(\frac{AF - CD}{2DF}\right)^2 + c_2 \left(\frac{AF - CD}{2DF}\right) + c_3 > 0\,. \tag{A.71}$$

After some manipulation, inequalities (A.70) and (A.71) reduce to $\theta_2^2/E^2 > 0$ and $\theta_2^2/(4DF) > 0$, respectively, which both hold in this case. We finally show that

$$R_{2_B} = \frac{-c_2 - \sqrt{\Delta_c}}{2c_1} < \frac{AF - CD}{DF}\,, \tag{A.72}$$

which completes the proof of (A.69). Inequality (A.72) reduces to

$$\sqrt{\Delta_c} < (E^2 - 4DF)(AF - CD)\,, \tag{A.73}$$

where both sides of (A.73) are positive. The inequality can therefore be squared and reduced to

$$-4DF\,(E^2 - 4DF)\,K > 0\,,$$

which holds in this case. Therefore, if (A.44) and (A.45) hold with $R_2 \neq (AE - BD)/(DE)$, all network elements are positive for both sets of solutions.

We now assume conditions (A.44) and (A.46) hold with $R_2 \neq (AE - BD)/(DE)$. From (A.46) it follows that $c_1 = c_2 = c_3 = 0$ in (A.47) and any value of x solves (A.47). Any value of $R_2 \neq (AE - BD)/(DE)$ strictly less than $(AF - CD)/(DF)$ still yields positive R_3 and R_1 (from (A.48) and (A.67), respectively). From (A.65) it follows that $\theta_2 = 0$, which, using (A.66), implies that

$$\frac{AE - BD}{DE} = \frac{AF - CD}{2DF}\,.$$

Therefore $R_2 \neq (AF - CD)/(2DF)$ and the expression for L_1 in (A.68), which is still valid in this case, thus reduces to

$$L_1 = \frac{2DF\,R_1R_3}{CE} \cdot \frac{\left(R_2 - \frac{AF-CD}{2DF}\right)}{\left(R_2 - \frac{AE-BD}{DE}\right)} = \frac{2DF\,R_1R_3}{CE}\,, \tag{A.74}$$

which ensures positivity of L_1 and hence L_2.

We now check positivity when conditions (A.44) and (A.45) hold in the case that $R_3 = B/E$, which, as noted before, is equivalent to

$$R_2 = \frac{AE - BD}{DE} \tag{A.75}$$

from (A.48). In this case (A.64) cannot be used to obtain L_1. However, we now have expressions for R_1, R_2, R_3, L_2, and k which only contain L_1, together with A, B, C, D, E, F. Substituting such expressions into (A.58) we get the quadratic equation (A.52) in L_1, from which we can conclude that L_1 is finite. Since (A.55) also needs to be satisfied (equivalently (A.64)), it follows that $R_1R_2 + R_2R_3 - R_3^2 = 0$. Considering the values obtained for R_1, R_2, and R_3 we get

$$R_1R_2 + R_2R_3 - R_3^2 = \frac{B^2\,\theta_2}{DE^2\,(BF - CE)} = 0\,, \tag{A.76}$$

from which we can conclude that $\theta_2 = 0$ in this case. From (A.49), $R_1 = BC/(BF - CE)$, which is always positive. It is easily seen that the first and third coefficients in (A.52) are positive, while the second is negative. The discriminant of (A.52),

$$\Delta_d = B^4E^2F^2\,(BF - CE)^2(E^2 - 4DF)\,,$$

is greater than zero, and hence the quadratic has two distinct positive roots. We finally note that substituting the expression for R_2 given in (A.75) into the left-hand side of (A.47), we obtain θ_2^2/E^2, which equals zero in this case. Therefore R_2 can still be computed as a root of (A.47).

The remaining case to be dealt with is when (A.44) and (A.46) hold and $R_2 = (AE - BD)/(DE)$. In this case the reasoning of the previous paragraph remains valid, and we notice that the discriminant of (A.52) $\Delta_d = 0$; hence there are two coincident solutions for L_1. □

Figure A.4 shows the realizability region for the network, plotted on the (U, V)-plane for $W = 0.5$ (i.e., $AF - CD > 0$). The polynomial θ_2 expressed in terms of U, V, W is $\theta_{2,c} = V(W^2 + 1) - 2UW$, while the expressions in terms of U, V, W for K and $E^2 - 4DF$ can be found in Table 6.10. We note that in the interior of the hatched region there are always two solutions to the realization problem, with no solutions on the boundary $K_c = 0$ except when $V = 1$ (i.e., $E^2 - 4DF = 0$), when there are infinitely many solutions. It can be shown that $V = 1$ and $K_c = 0$ imply $\theta_{2,c} = 0$.

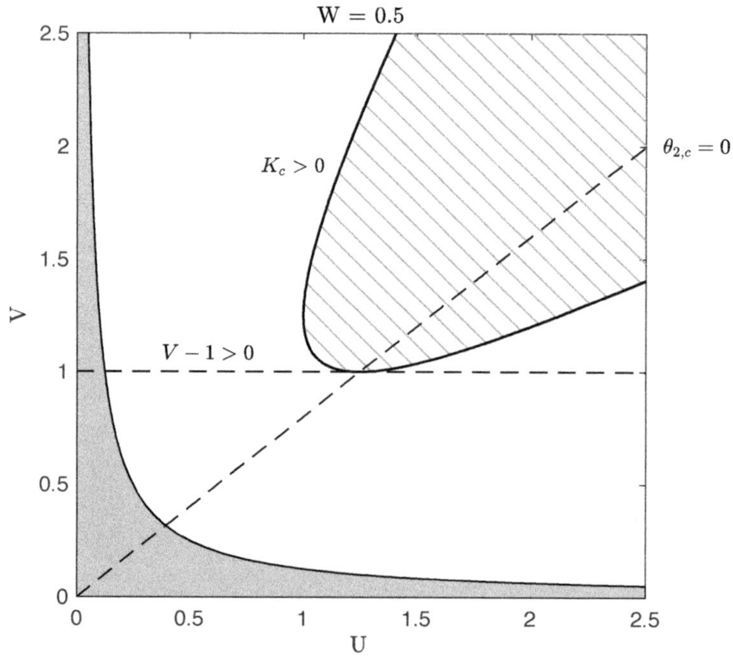

Figure A.4. *Realizability region for the network on the (U, V)-plane for $W = 0.5$. The interior of the hatched region corresponds to case* (A.45) *(with $R_2 = (AE - BD)/(DE)$ being a repeated root in* (A.47) *when $\theta_{2,c} = 0$), while the point where the curves K_c and $V - 1$ intersect corresponds to case* (A.46).

A.4 ▪ Equivalence class V_C^1

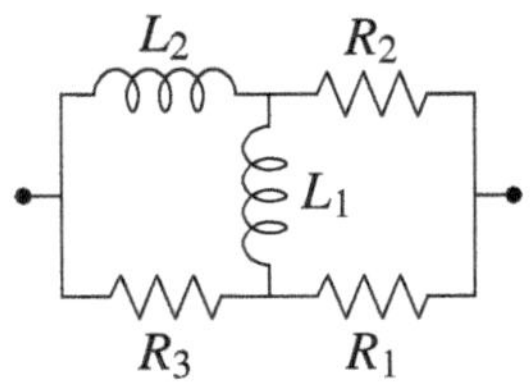

Figure A.5. *Network #60 from subfamily V_C. By Lemma* 3 *in* [47] *it can only realize regular impedances.*

Theorem A.4. *The positive-real biquadratic impedance* (3.1) *with A, B, C, D, E, $F > 0$ can be realized as in Figure* A.5, *with R_1, R_2, R_3, L_1, and L_2 positive and finite, if and only if*

$$AF - CD > 0\,, \tag{A.77}$$

$$\eta \leq 0\,, \tag{A.78}$$

where η is defined in Table 6.10. *If conditions* (A.77)–(A.78) *are satisfied, then R_1 is either of the two positive roots of the quadratic equation in x*

$$DF\,\tau_1\,x^2 - \psi\,x + AC\,\tau_2 = 0\,, \tag{A.79}$$

where τ_1, τ_2, and ψ are defined in Table 6.10, *and*

$$R_2 = \frac{CR_1}{FR_1 - C}\,, \tag{A.80}$$

$$R_3 = \frac{A - DR_1}{D}\,, \tag{A.81}$$

$$L_1 = \frac{R_1^2\,(-DEF\,R_1 + \gamma_2)}{(AF + CD)(FR_1 - C)}\,, \tag{A.82}$$

$$L_2 = \frac{R_1\,\gamma_1 - ABC}{(AF + CD)(FR_1 - C)}\,, \tag{A.83}$$

where

$$\gamma_1 = A(BF - CE) + BCD\,, \tag{A.84}$$

$$\gamma_2 = F(AE - BD) + CDE\,. \tag{A.85}$$

Proof. *Necessity.* The impedance of the network shown in Figure A.5 is a biquadratic, which can be computed as

$$Z(s) = \frac{n(s)}{d(s)}\,, \tag{A.86}$$

where

$$\begin{aligned} n(s) &= L_1L_2(R_1 + R_3)s^2 + \big(L_2(R_1R_2 + R_1R_3 + R_2R_3) + R_2L_1(R_1 + R_3)\big)s \\ &\quad + R_1R_2R_3\,, \\ d(s) &= L_1L_2s^2 + \big(L_2(R_1 + R_2) + L_1(R_1 + R_2 + R_3)\big)s + R_3(R_1 + R_2)\,. \end{aligned}$$

Equating impedance (A.86) with (3.1), we obtain

$$L_1L_2(R_1+R_3) = kA\,, \tag{A.87}$$
$$L_2(R_1R_2+R_1R_3+R_2R_3)+R_2L_1(R_1+R_3) = kB\,, \tag{A.88}$$
$$R_1R_2R_3 = kC\,, \tag{A.89}$$
$$L_1L_2 = kD\,, \tag{A.90}$$
$$L_2(R_1+R_2)+L_1(R_1+R_2+R_3) = kE\,, \tag{A.91}$$
$$R_3(R_1+R_2) = kF\,, \tag{A.92}$$

where k is a positive constant. It can be calculated that

$$AF-CD = k^{-2}\,L_1L_2R_3(R_1^2+R_1R_3+R_2R_3)\,,$$
$$\eta = -k^{-4}\,L_1L_2R_3^2[L_1R_2(R_1+R_3)-L_2R_1(R_1+R_2)]^2\,,$$

and hence (A.77)–(A.78) are necessary.

Sufficiency. Given a positive-real impedance (3.1) with $A, B, C, D, E, F > 0$ satisfying conditions (A.77)–(A.78), we now show that we can find R_1, R_2, R_3, L_1, L_2 positive which satisfy (A.87)–(A.92) with $k > 0$. From (A.89) we obtain

$$k = \frac{R_1R_2R_3}{C}\,. \tag{A.93}$$

Eliminating L_1L_2 from (A.87) and (A.90) we obtain (A.81), and substituting (A.93) into (A.92) we get (A.80). Substituting (A.93), (A.80), and (A.81) into (A.88) and (A.91), and solving the equations for L_1 and L_2, we obtain (A.82) and (A.83). We now have expressions for elements R_2, R_3, L_1, L_2, and k which only contain R_1, together with A, B, C, D, E, and F. Substituting such expressions into (A.87) we obtain the quadratic equation (A.79). The discriminant of (A.79),

$$\Delta = (AF+CD)^2(\eta-4ACDF)\,\eta\,, \tag{A.94}$$

is greater than or equal to zero, by virtue of (A.78). It can be easily seen that the first and third coefficients in (A.79) are negative, since $\tau_1 = \eta - B^2DF$ and $\tau_2 = \eta - ACE^2$, while the second coefficient is positive, since $-\psi = -(AF+CD)\eta + 2ABCDEF$. Therefore the equation has two real positive roots, which are distinct if the discriminant Δ is strictly greater than zero.

It can be easily seen from the expressions in Table 6.10 that $K = \eta - 4ACDF$, and hence (A.78) implies $K < 0$. From $K < 0$ and (A.77) it follows that

$$BF-CE > 0\,, \tag{A.95}$$
$$AE-BD > 0\,, \tag{A.96}$$

and therefore γ_1 and γ_2 in (A.84)–(A.85) are positive. We denote the two solutions of (A.79) as R_{1_A} and R_{1_B}, with $R_{1_A} \le R_{1_B}$. From (A.80) and (A.81), R_2 and R_3 are positive for both solutions, providing

$$C/F < R_{1_A} \le R_{1_B} < A/D \tag{A.97}$$

and, from (A.82) and (A.83) (assuming (A.97) holds), L_1 and L_2 are positive if

$$ABC/\gamma_1 < R_{1_A} \le R_{1_B} < \gamma_2/(DEF)\,. \tag{A.98}$$

It can be easily verified that $C/F < ABC/\gamma_1$ and $\gamma_2/(DEF) < A/D$; hence if inequality (A.98) holds, so does (A.97). Since R_{1_A} and R_{1_B} are the solutions to (A.79), which represents a parabola that opens down, inequality (A.98) holds if and only if ABC/γ_1 and $\gamma_2/(DEF)$ are located, respectively, to the left and to the right of the parabola's axis of symmetry and (A.79) evaluated at ABC/γ_1 and $\gamma_2/(DEF)$ gives negative values, that is,

$$\frac{ABC}{\gamma_1} < \frac{\psi}{2DF\,\tau_1} < \frac{\gamma_2}{DEF}\,, \tag{A.99}$$

and

$$DF\,\tau_1\left(\frac{ABC}{\gamma_1}\right)^2 - \psi\left(\frac{ABC}{\gamma_1}\right) + AC\,\tau_2 < 0\,, \tag{A.100}$$

$$DF\,\tau_1\left(\frac{\gamma_2}{DEF}\right)^2 - \psi\left(\frac{\gamma_2}{DEF}\right) + AC\,\tau_2 < 0\,. \tag{A.101}$$

The left and right inequalities in (A.99) can be rewritten as

$$\begin{aligned} 2ABCDF\,\tau_1 - \psi\gamma_1 &= (AF+CD)(2ABCDF(AF+CD) - \eta\gamma_1) > 0\,,\\ DEF\,\psi - 2DF\,\tau_1\gamma_2 &= DF(AF+CD)(2BDF(AF+CD) - E\,\eta) > 0\,, \end{aligned}$$

which both hold. After some manipulation, inequalities (A.100) and (A.101), respectively, reduce to

$$\begin{aligned} -A^2C^2(AE-BD)(BF-CE)(AF+CD)^2/\gamma_1^2 < 0\,,\\ -(AE-BD)(BF-CE)(AF+CD)^2/E^2 < 0\,, \end{aligned}$$

which always hold. Therefore the values of the five elements and the constant k are all positive for both sets of solutions. □

A.5 ▪ Equivalence class V_D^1

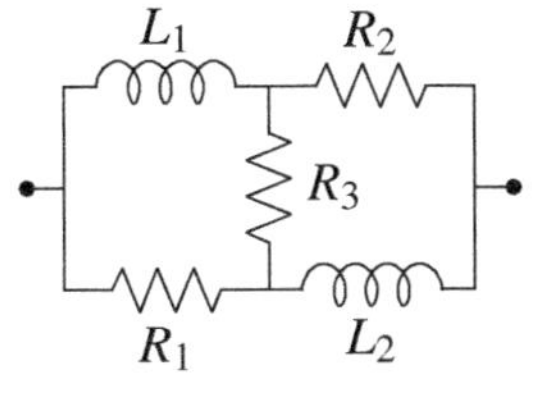

Figure A.6. *Network #61 from subfamily V_D. By Lemma 3 in [47] it can only realize regular impedances.*

Theorem A.5. *The positive-real biquadratic impedance (3.1) with A, B, C, D, E, $F > 0$ can be realized as in Figure A.6, with R_1, R_2, R_3, L_1, and L_2 positive and finite, if and only if*

$$K \le 0\,, \tag{A.102}$$

$$\mu_1 \le 0\,, \tag{A.103}$$

$$AF - CD > 0\,, \tag{A.104}$$

where K is defined in (3.3) *and μ_1 in Table* 6.10.[4]

If conditions (A.102)–(A.104) *are satisfied, then R_3 is either of the two positive roots of the quadratic equation in x*

$$-D\,\lambda_1\,x^2 + \rho_1\,x - C\,\lambda_2 = 0\,, \tag{A.105}$$

where λ_1, λ_2, and ρ_1 are defined in Table 6.10, *R_2 is either of the two positive roots of the quadratic equation in y*

$$d_1\,y^2 + d_2\,y + d_3 = 0\,, \tag{A.106}$$

where

$$d_1 = D(C - FR_3)\,, \tag{A.107}$$
$$d_2 = -(C - FR_3)(A - DR_3)\,, \tag{A.108}$$
$$d_3 = -CR_3(A - DR_3), \tag{A.109}$$

and R_1 is the other root. The values of the inductors are

$$L_1 = \frac{-D(R_1 + R_3)(R_1 - R_2)}{B - ER_1}\,, \tag{A.110}$$
$$L_2 = \frac{D(R_2 + R_3)(R_1 - R_2)}{B - ER_2} \tag{A.111}$$

if $R_1 \neq R_2$. If $R_1 = R_2$, then L_1 and L_2 are the two positive roots of the quadratic equation in z

$$p_1\,z^2 + p_2\,z + p_3 = 0\,, \tag{A.112}$$

where

$$p_1 = E^2F(AE - BD)\,, \tag{A.113}$$
$$p_2 = -BE^2(2AE - 3BD)\,, \tag{A.114}$$
$$p_3 = B(2AE - 3BD)(AE - BD)\,. \tag{A.115}$$

Proof. *Necessity.* The impedance of the network shown in Figure A.6 is a biquadratic, which can be computed as

$$Z(s) = \frac{n(s)}{d(s)}\,, \tag{A.116}$$

where

$$n(s) = L_1L_2(R_1 + R_2 + R_3)\,s^2 + \big(L_1R_1(R_2 + R_3) + L_2R_2(R_1 + R_3)\big)s + R_1R_2R_3\,,$$
$$d(s) = L_1L_2\,s^2 + \big(L_1(R_2 + R_3) + L_2(R_1 + R_3)\big)s + R_1R_2 + R_1R_3 + R_2R_3\,.$$

[4]Conditions (A.102)–(A.104) could be rewritten with $AF - CD > 0$ replaced by $AF - 9CD \geq 0$. Even though $AF - CD > 0$ is apparently a weaker condition, it is actually equivalent in the theorem statement, since conditions (A.102)–(A.103) imply $AF - 9CD \geq 0$ when $AF - CD > 0$, as shown in [18, Theorem 2].

Equating impedance (A.116) with (3.1), we obtain

$$L_1L_2(R_1+R_2+R_3)=kA\,, \tag{A.117}$$
$$L_1R_1(R_2+R_3)+L_2R_2(R_1+R_3)=kB\,, \tag{A.118}$$
$$R_1R_2R_3=kC\,, \tag{A.119}$$
$$L_1L_2=kD\,, \tag{A.120}$$
$$L_1(R_2+R_3)+L_2(R_1+R_3)=kE\,, \tag{A.121}$$
$$R_1R_2+R_1R_3+R_2R_3=kF\,, \tag{A.122}$$

where k is a positive constant. It can be calculated that

$$K=-k^{-4}\,L_1L_2\big[L_1R_1(R_2+R_3)^2-L_2R_2(R_1+R_3)^2\big]^2\,, \tag{A.123}$$
$$\mu_1=-k^{-4}\,L_1L_2\big[L_1R_1(R_2^2-R_3^2)-L_2R_2(R_1^2-R_3^2)\big]^2\,, \tag{A.124}$$
$$AF-CD=k^{-2}\,L_1L_2(R_1+R_2)(R_1R_2+R_1R_3+R_2R_3+R_3^2)\,, \tag{A.125}$$

and hence (A.102)–(A.104) are necessary.

Sufficiency. Given a positive-real impedance (3.1) with $A, B, C, D, E, F>0$ satisfying conditions (A.102)–(A.104), we now show that we can find R_1, R_2, R_3, L_1, L_2 positive which satisfy (A.117)–(A.122) with $k>0$. Solving (A.118) and (A.121) for L_1 and L_2, we obtain

$$L_1=\frac{k(B-ER_2)}{(R_2+R_3)(R_1-R_2)}\,, \tag{A.126}$$
$$L_2=\frac{-k(B-ER_1)}{(R_1+R_3)(R_1-R_2)}\,, \tag{A.127}$$

assuming $R_1\neq R_2$ and $R_1,\ R_2\neq B/E$, and from (A.120), (A.126)–(A.127) we obtain the expressions for L_1 and L_2 in (A.110) and (A.111). Eliminating L_1L_2 from (A.117) and (A.120) we obtain

$$R_1=A/D-R_2-R_3\,, \tag{A.128}$$

and from (A.122) and (A.128) we get

$$k=\frac{(A/D-R_2-R_3)(R_2+R_3)+R_2R_3}{F}\,. \tag{A.129}$$

Substituting (A.128) and (A.129) into (A.119), we get the quadratic equation (A.106), with $y=R_2$, from which an expression for R_2^2 can be found, as follows:

$$R_2^2=\frac{(A-DR_3)(C(R_2+R_3)-FR_2R_3)}{D(C-FR_3)}\,. \tag{A.130}$$

We now have expressions for elements R_1, L_1, L_2 and for the constant k which only contain R_2 and R_3, together with A, B, C, D, E, and F, and an expression for R_2^2 which contains R_3, together with the same six coefficients, and is linear in R_2. Substituting the expressions for L_1, L_2, and k into (A.120) we obtain a polynomial in R_2 and R_3 of fourth order. Then, substituting for R_2^2 from (A.130) twice to reduce the power of R_2, after further manipulation all terms in R_2 cancel out, and we obtain the quadratic equation (A.105) in R_3. The discriminant of (A.105),

$$\Delta_c=K\mu_1\,, \tag{A.131}$$

is greater than or equal to zero, by virtue of (A.102)–(A.103). Conditions (A.102) and (A.104) imply that $\rho_1 > 0$ and

$$BF - CE > 0\,,$$
$$AE - BD > 0\,.$$

Hence, using again (A.102) and (A.104), λ_1 and λ_2 are both positive (note this also follows from [47, Lemma 7]). Therefore (A.105) has two real positive roots

$$R_{3_A} = \frac{\rho_1 - \sqrt{\Delta_c}}{2D\lambda_1}\,, \tag{A.132}$$

$$R_{3_B} = \frac{\rho_1 + \sqrt{\Delta_c}}{2D\lambda_1}\,, \tag{A.133}$$

where $R_{3_A} \leq R_{3_B}$, which coincide if and only if $\Delta_c = 0$, i.e., if K and/or μ_1 are zero.

If $R_1 = R_2$, then (A.110) and (A.111) do not apply. In this case the left-hand sides of (A.118) and (A.121) are multiples of $L_1 + L_2$, which can be eliminated to yield $R_1 = R_2 = B/E$. As before, (A.128) holds and it follows that

$$R_3 = \frac{AE - 2BD}{DE}\,. \tag{A.134}$$

Solving (A.122) yields $k = B(2AE - 3BD)/(DE^2F)$, which implies that

$$\beta := BF(AE - 2BD) - CE(2AE - 3BD) \tag{A.135}$$

is zero, using (A.119). It can then be verified that R_3 still satisfies (A.105) (and is therefore positive) and B/E is a double root of (A.106). From (A.134) we can therefore conclude that, in this case,

$$AE - 2BD > 0\,. \tag{A.136}$$

Solving for L_1 or L_2 from (A.120) and (A.121) shows that they are both solutions of the quadratic (A.112).

Positivity of R_1 and R_2

From (A.117) and (A.120), and from (A.119) and (A.122), respectively, it can be easily shown that

$$R_1 + R_2 + R_3 = A/D\,,$$
$$\left(\frac{1}{R_1} + \frac{1}{R_2} + \frac{1}{R_3}\right)^{-1} = C/F\,,$$

and hence each R_i is smaller than A/D and greater than C/F. It follows that coefficients d_1 and d_3 in (A.106) are negative, and d_2 is positive. Hence, (A.106) has two positive solutions if and only if the discriminant

$$\Delta_d = (A - DR_3)(C - FR_3)\left[DF\,R_3^2 - (AF - 3CD)\,R_3 + AC\right] \tag{A.137}$$

is nonnegative, i.e., if and only if

$$DF\,R_3^2 - (AF - 3CD)\,R_3 + AC \leq 0\,. \tag{A.138}$$

Substituting (A.132) into (A.138), after some manipulation (A.138) is equivalent to

$$2D^2(BF - CE)(BF - 3CE)\sqrt{\Delta_c} \leq -2D^2\xi\,, \tag{A.139}$$

where

$$\xi = K(BF - 3CE)^2 - 2C\lambda_1\beta \tag{A.140}$$
$$= \mu_1 (BF - CE)^2 + 2C\lambda_1\beta\,, \tag{A.141}$$

with β defined as in (A.135). From (A.140) it follows that if $\beta \geq 0$, then $\xi \leq 0$, and from (A.141) it follows that if $\beta < 0$, then $\xi < 0$. Therefore we can conclude that $\xi \leq 0$. Hence, if $BF - 3CE \leq 0$, inequality (A.139) holds. If $BF - 3CE > 0$, both sides of (A.139) are nonnegative and the inequality can be squared and reduced to

$$-16C^2D^4\beta^2\lambda_1^2 \leq 0\,, \tag{A.142}$$

which holds. A similar argument holds if we substitute (A.133) into (A.138). Therefore, considering either R_{3_A} or R_{3_B}, all solutions to (A.106) are positive. The two positive values of R_2 thus obtained from R_{3_A} are

$$R'_{2_A},\ R''_{2_A} = \left.\frac{-d_2 \pm \sqrt{\Delta_d}}{2d_1}\right|_{R_3=R_{3_A}}.$$

The same expression holds for the other two solutions R'_{2_B} and R''_{2_B} with R_{3_A} replaced by R_{3_B}. From (A.106) the sum of the two roots is

$$R'_{2_A} + R''_{2_A} = A/D - R_{3_A}\,,$$

and hence from (A.128) it follows that

$$R'_{1_A} = A/D - R'_{2_A} - R_{3_A} = R''_{2_A}\,, \tag{A.143}$$
$$R''_{1_A} = A/D - R''_{2_A} - R_{3_A} = R'_{2_A}\,. \tag{A.144}$$

Similar equalities hold for R'_{1_B} and R''_{1_B}. This is due to the symmetry of the network and can be seen from (A.117)–(A.122): if the resistors R_1, R_2 and the inductors L_1, L_2 are swapped, the impedance is unchanged. Note we have shown that R_1 and R_2 are the two positive solutions of (A.106) for either choice of R_3.

Positivity of L_1 and L_2

When $R_1 \neq R_2$, it follows from (A.110)–(A.111) that the two following cases result in positive values for both inductances:

$$R_2 < B/E < R_1\,, \tag{A.145}$$
$$R_1 < B/E < R_2\,. \tag{A.146}$$

Since R_1 and R_2 are the two roots of the same quadratic equation, only one of the two cases can occur depending on whether $R_1 \gtrless R_2$, which is a matter of how the two roots of (A.106) are labeled. We will now show that, considering the set of solutions R_{3_A}, R'_{2_A}, and R'_{1_A}, with $R'_{2_A} < R''_{2_A}$, (A.145) always holds if conditions (A.102)–(A.104) hold.

Since R'_{2_A} and $R'_{1_A} = R''_{2_A}$ are the solutions to the quadratic equation (A.106), which represents a parabola that opens down, inequality (A.145) holds if and only if (A.106) evaluated at B/E gives a positive value, that is,

$$d_1\, B^2/E^2 + d_2\, B/E + d_3 > 0\,. \tag{A.147}$$

After some manipulation, inequality (A.147) can be reduced to

$$\gamma_1 R_{3_A}^2 + \gamma_2 R_{3_A} + \gamma_3 < 0\,, \tag{A.148}$$

where

$$\begin{aligned}\gamma_1 &= DE(BF - CE)\,,\\ \gamma_2 &= -(AE - BD)(BF - CE)\,,\\ \gamma_3 &= BC(AE - BD)\,.\end{aligned}$$

Substituting (A.132) into (A.148), after some manipulation the inequality reduces to

$$2D^2(BF - CE)(BF - 3CE)(AF - CD)\sqrt{\Delta_c} < -2D^2(AF - CD)\,\xi\,,$$

which always holds, since (A.139) holds with strict inequality.

For the case $R_1 = R_2$, as shown by (A.136), $AE - 2BD > 0$, and hence also $2AE - 3BD > 0$, which means that p_1 and p_3 in (A.112) are positive while p_2 is negative. Using the fact that $\beta = 0$, the discriminant of (A.112) simplifies to the following expression:

$$\Delta_p = -BE^3\,(2AE - 3BD)^2\,\big[2(AF - CD) - BE\big]\,.$$

It can be verified that the following identity always holds:

$$2\,K = (AF - CD)\big[2(AF - CD) - BE\big] - \beta\,, \tag{A.149}$$

which implies that $2(AF - CD) - BE \leq 0$ from (A.102), (A.104) and $\beta = 0$. It follows that $\Delta_p \geq 0$ and that both solutions of (A.112) are positive. □

A.6 ▪ Equivalence class V_E^1

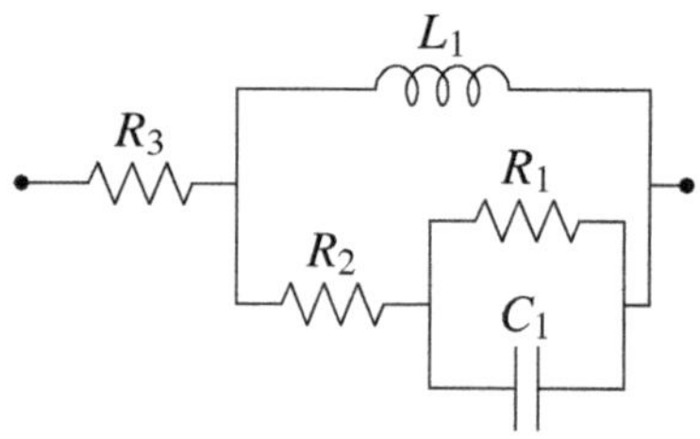

Figure A.7. *Network #69 from subfamily V_E. By Theorem* 1 *in* [47] *it can only realize regular impedances.*

Theorem A.6. *The positive-real biquadratic impedance* (3.1) *with A, B, C, D, E, $F > 0$ can be realized as in Figure* A.7, *with R_1, R_2, R_3, C_1, and L_1 positive and finite, if and only if*

$$AF - CD > 0\,, \tag{A.150}$$

$$\lambda_1 > 0\,, \tag{A.151}$$

$$K > 0\,, \tag{A.152}$$

where λ_1 *and* K *are defined in Table* 6.10. *If conditions* (A.150)–(A.152) *are satisfied, then*

$$R_1 = \frac{K}{D\lambda_1}, \tag{A.153}$$

$$R_2 = \frac{AF - CD}{DF}, \tag{A.154}$$

$$R_3 = \frac{C}{F}, \tag{A.155}$$

$$C_1 = \frac{D^2(BF - CE)}{K}, \tag{A.156}$$

$$L_1 = \frac{BF - CE}{F^2}. \tag{A.157}$$

Proof. Note this result is given without proof in [47, Appendix A], with R_1 and R_3 interchanged. We provide here a proof for convenience.

Necessity. The impedance of the network shown in Figure A.7 is a biquadratic, which can be computed as

$$Z(s) = \frac{n(s)}{d(s)}, \tag{A.158}$$

where

$$\begin{aligned} n(s) &= R_1L_1C_1(R_2 + R_3)s^2 + \big(L_1(R_1 + R_2 + R_3) + R_1R_2R_3C_1\big)s \\ &\quad + R_3(R_1 + R_2), \\ d(s) &= R_1L_1C_1s^2 + \big(L_1 + R_1R_2C_1\big)s + R_1 + R_2. \end{aligned}$$

Equating impedance (A.33) with (3.1), we obtain, for a positive constant k,

$$R_1L_1C_1(R_2 + R_3) = kA, \tag{A.159}$$

$$L_1(R_1 + R_2 + R_3) + R_1R_2R_3C_1 = kB, \tag{A.160}$$

$$R_3(R_1 + R_2) = kC, \tag{A.161}$$

$$R_1L_1C_1 = kD, \tag{A.162}$$

$$L_1 + R_1R_2C_1 = kE, \tag{A.163}$$

$$R_1 + R_2 = kF, \tag{A.164}$$

which are a set of necessary and sufficient conditions for (3.1) to be realized as in Figure A.7. It can be calculated that

$$AF - CD = k^{-2}\, R_1R_2L_1C_1\,(R_1 + R_2) > 0, \tag{A.165}$$

$$\lambda_1 = k^{-3}\, L_1^2\,(R_1 + R_2)^2 > 0, \tag{A.166}$$

$$K = k^{-4}\, R_1^2L_1^3C_1\,(R_1 + R_2)^2 > 0, \tag{A.167}$$

and hence (A.150)–(A.152) are necessary.

Sufficiency. Given a positive-real impedance (3.1) with $A, B, C, D, E, F > 0$ satisfying conditions (A.150)–(A.152), we now show that we can find R_1, R_2, R_3, L_1, C_1 positive which satisfy (A.159)–(A.164) with $k > 0$.

From (A.164) we obtain $k = (R_1 + R_2)/F$ and eliminating the term $R_1 + R_2$ from (A.161) and (A.164) we get (A.155). Eliminating the term $R_1 L_1 C_1$ from (A.159) and (A.162) and solving for R_2 we obtain (A.154), while eliminating C_1 from (A.160) and (A.163) we get (A.157). Solving for C_1 from (A.162) using expressions obtained we find

$$C_1 = \frac{AF - CD + DFR_1}{R_1\,(BF - CE)} \,. \tag{A.168}$$

We now have expressions for R_2, R_3, L_1, C_1, and k which only contain R_1 together with A, B, C, D, E, F. Substituting such expressions into (A.163) and solving for R_1 we obtain (A.153), and substituting the latter into (A.168), we get (A.156).

Conditions (A.150) and (A.151) imply $BF - CE > 0$ from the expression for λ_1. Hence all network elements are positive. □

A.7 ▪ Equivalence class V_F^1

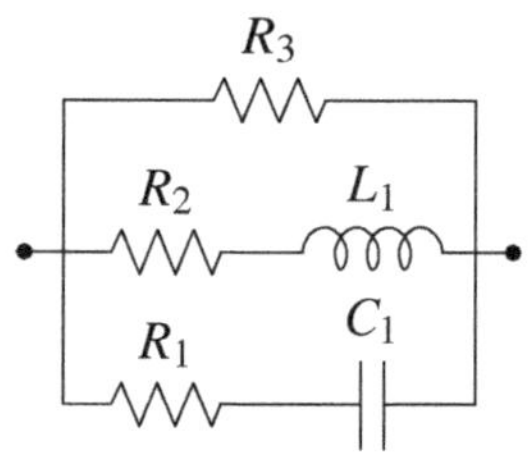

Figure A.8. *Network #101 from subfamily V_F. By Theorem 1 in [47] it can only realize regular impedances.*

Theorem A.7. *The positive-real biquadratic impedance (3.1) with A, B, C, D, E, $F > 0$ can be realized as in Figure A.8, with R_1, R_2, R_3, L_1, and C_1 positive and finite, if and only if*

$$K > 0\,, \tag{A.169}$$

$$\tau_1 < 0\,, \tag{A.170}$$

where K is defined in (3.3) and τ_1 in Table 6.10, or

$$K = \tau_1 = 0\,. \tag{A.171}$$

If (A.169)–(A.170) hold, then R_3 is the largest positive root of the quadratic equation in x

$$\tau_1\, x^2 + (AF + CD)(B^2 - 4AC)\, x - AC(B^2 - 4AC) = 0\,, \tag{A.172}$$

while if (A.171) holds, then R_3 may take any value such that

$$R_3 > \max\left\{\frac{A}{D}, \frac{B}{E}, \frac{C}{F}\right\} . \tag{A.173}$$

The other network elements are given by

$$R_1 = \frac{AR_3}{DR_3 - A}, \tag{A.174}$$

$$R_2 = \frac{CR_3}{FR_3 - C}, \tag{A.175}$$

$$L_1 = \frac{DR_3^2(R_1 + R_2)}{(R_1 + R_3)(ER_3 - B)}, \tag{A.176}$$

$$C_1 = \frac{D(R_2 + R_3)}{FL_1(R_1 + R_3)}. \tag{A.177}$$

Proof. *Necessity.* The impedance of the network shown in Figure A.8 is a biquadratic, which can be computed as

$$Z(s) = \frac{n(s)}{d(s)}, \tag{A.178}$$

where

$$n(s) = R_1R_3L_1C_1\,s^2 + R_3(R_1R_2C_1 + L_1)s + R_2R_3\,,$$
$$d(s) = L_1C_1(R_1 + R_3)\,s^2 + \big(C_1(R_1R_2 + R_1R_3 + R_2R_3) + L_1\big)s + R_2 + R_3\,.$$

Equating impedance (A.178) with (3.1) we obtain, for a positive constant k,

$$R_1R_3L_1C_1 = kA\,, \tag{A.179}$$

$$R_3(R_1R_2C_1 + L_1) = kB\,, \tag{A.180}$$

$$R_2R_3 = kC\,, \tag{A.181}$$

$$L_1C_1(R_1 + R_3) = kD\,, \tag{A.182}$$

$$C_1(R_1R_2 + R_1R_3 + R_2R_3) + L_1 = kE\,, \tag{A.183}$$

$$R_2 + R_3 = kF\,, \tag{A.184}$$

which are a set of necessary and sufficient conditions for (3.1) to be realized as in Figure A.8. It can be calculated that

$$K = k^{-4}\,C_1L_1R_3^4\,(C_1R_1R_2 - L_1)^2 \geq 0\,, \tag{A.185}$$

$$\tau_1 = -k^{-4}\,C_1L_1R_3^2\,(R_1R_2 + R_1R_3 + R_2R_3)\,(C_1R_1R_2 - L_1)^2 \leq 0\,, \tag{A.186}$$

and hence

$$K = 0 \;\Leftrightarrow\; \tau_1 = 0 \;\Leftrightarrow\; L_1 = C_1R_1R_2$$

and (A.169)–(A.171) are necessary.

Sufficiency. Given a positive-real impedance (3.1) with A, B, C, D, E, $F > 0$, satisfying either conditions (A.169)–(A.170) or condition (A.171), we now show that we can find R_1, R_2, R_3, L_1, C_1 positive which satisfy (A.179)–(A.184) with $k > 0$.

From (A.184) we obtain

$$k = \frac{R_2 + R_3}{F}, \tag{A.187}$$

and substituting (A.187) into (A.182) we get (A.177). Equations (A.180) and (A.183) now reduce to

$$
\begin{aligned}
FR_3(R_1+R_3)\,L_1^2 - B(R_1+R_3)(R_2+R_3)\,L_1 + DR_1R_2R_3(R_2+R_3) &= 0\,,\\
F(R_1+R_3)\,L_1^2 - E(R_1+R_3)(R_2+R_3)\,L_1 + D(R_2+R_3)(R_1R_2 &\\
+ R_1R_3 + R_2R_3) &= 0\,.
\end{aligned}
$$

Eliminating the term in L_1^2 from the two equations and solving for L_1 we obtain (A.176). Eliminating L_1C_1 from (A.179) and (A.182) and solving for R_1 we obtain (A.174), while from (A.181) and (A.184) we obtain (A.175). We now have expressions for R_1, R_2, L_1, C_1, and k which only contain R_3 together with A, B, C, D, E, F. Substituting such expressions into (A.180) we obtain the quadratic equation (A.172).

If conditions (A.169)–(A.170) hold, then from the expression for τ_1 we have that $DF(B^2-4AC) > K > 0$, and hence the first and third coefficients in (A.172) are negative while the second is positive. The discriminant of (A.172) is

$$\Delta = (B^2-4AC)(ABF-2ACE+BCD)^2\,.$$

It can be verified that the following identity always holds:

$$(ABF-2ACE+BCD)^2 = 4AC\,K + (B^2-4AC)(AF-CD)^2\,,$$

and hence $ABF-2ACE+BCD \neq 0$ in this case. Therefore $\Delta > 0$ and the quadratic has two distinct positive roots.

From (A.174), (A.175), and (A.176), respectively, we have that R_1, R_2, and L_1 are positive if

$$R_3 > A/D\,, \tag{A.188}$$

$$R_3 > C/F\,, \tag{A.189}$$

$$R_3 > B/E\,. \tag{A.190}$$

Since (A.172) represents a parabola that opens down, (A.188), (A.189), and (A.190) hold for the largest solution of (A.172) if (A.172) evaluated at A/D, C/F, and B/E gives positive values, that is,

$$\tau_1\,A^2/D^2 + (AF+CD)(B^2-4AC)\,A/D - AC(B^2-4AC) > 0\,, \tag{A.191}$$

$$\tau_1\,C^2/F^2 + (AF+CD)(B^2-4AC)\,C/F - AC(B^2-4AC) > 0\,, \tag{A.192}$$

$$\tau_1\,B^2/E^2 + (AF+CD)(B^2-4AC)\,B/E - AC(B^2-4AC) > 0\,. \tag{A.193}$$

After some manipulation, inequalities (A.191)–(A.193) reduce to

$$\frac{A^2\,K}{D^2} > 0\,, \qquad \frac{C^2\,K}{F^2} > 0\,, \qquad \frac{(ABF-2ACE+BCD)^2}{E^2} > 0\,,$$

which all hold in this case. This means that A/D, C/F, and B/E always lie between the two positive roots of (A.172), and so (A.188)–(A.190) are satisfied by the largest root only.

Finally, if condition (A.171) holds, then $B^2-4AC = 0$ (from the expression for τ_1) and any value of x solves (A.172). Any positive value of R_3 which satisfies (A.173) guarantees positivity of R_1, R_2, and L_1, from (A.174), (A.175), and (A.176), and the theorem statement follows. □

A.8 ▪ Equivalence class V_G^1

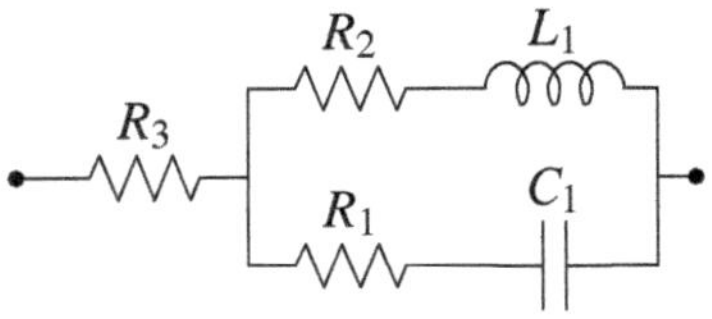

Figure A.9. *Network #104 from subfamily V_G. By Theorem 1 in [47] it can only realize regular impedances.*

Theorem A.8. *The positive-real biquadratic impedance (3.1) with A, B, C, D, E, $F > 0$ can be realized as in Figure A.9, with R_1, R_2, R_3, L_1, and C_1 positive and finite, if and only if the conditions of one of the following five cases are satisfied:*

Case (1) $AF - CD > 0$ *and one of*

(a) $\tau_1 < 0, \ \lambda_1 = 0\,,$ (A.194)

(b) $\lambda_1 > 0, \ \tau_1 = 0, \ \delta > 0\,,$ (A.195)

(c) $\tau_1 < 0, \ \lambda_1 > 0, \ K \geq 0, \ \delta > 0, \ \zeta_1 > 0\,.$ (A.196)

Case (2) $AF - CD \geq 0$ *and* $\tau_1\lambda_1 > 0\,.$ (A.197)

Case (3) $AF - CD = 0$ *and* $K = 0\,.$ (A.198)

Case (4) $AF - CD < 0$ *and one of*

(a) $\tau_1 < 0, \ \lambda_3 = 0\,,$ (A.199)

(b) $\lambda_3 > 0, \ \tau_1 = 0, \ \delta > 0\,,$ (A.200)

(c) $\tau_1 < 0, \ \lambda_3 > 0, \ K \geq 0, \ \delta > 0, \ \zeta_3 > 0\,.$ (A.201)

Case (5) $AF - CD \leq 0$ *and* $\tau_1\lambda_3 > 0\,.$ (A.202)

Polynomials K, λ_1, λ_3, ζ_1, ζ_3, δ, and τ_1 are defined in Table 6.10.

When the conditions in Case (1)(a), (1)(b), (2), (4)(a), (4)(b), *or* (5) *hold, R_3 is the smallest, positive root of the quadratic equation in x*

$$\gamma x^2 + 2DF\,\delta\, x + \tau_1 = 0\,, \quad \text{(A.203)}$$

where

$$\gamma = -DF\left(E^2 - 4DF\right)\,. \quad \text{(A.204)}$$

When the conditions in Case (1)(c) *or* (4)(c) *hold, R_3 is either of the two positive roots of the same quadratic equation, and when the conditions of Case* (3) *hold, R_3 may take any value such that $R_3 < A/D = C/F$. The other elements of the network can be computed as*

$$R_1 = -R_3 + A/D\,, \quad \text{(A.205)}$$

$$R_2 = -R_3 + C/F\,, \quad \text{(A.206)}$$

$$L_1 = D(R_1 + R_2)/E\,, \quad \text{(A.207)}$$

$$C_1 = \frac{E}{F(R_1 + R_2)}\,. \quad \text{(A.208)}$$

Proof. *Necessity.* The impedance of the network shown in Figure A.9 is a biquadratic, which can be computed as

$$Z(s) = \frac{n(s)}{d(s)}, \tag{A.209}$$

where

$$n(s) = L_1C_1(R_1 + R_3)s^2 + \big(L_1 + C_1(R_1R_2 + R_1R_3 + R_2R_3)\big)s + R_2 + R_3,$$
$$d(s) = L_1C_1s^2 + C_1(R_1 + R_2)s + 1.$$

Equating impedance (A.209) with (3.1), we obtain

$$L_1C_1(R_1 + R_3) = kA, \tag{A.210}$$
$$L_1 + C_1(R_1R_2 + R_1R_3 + R_2R_3) = kB, \tag{A.211}$$
$$R_2 + R_3 = kC, \tag{A.212}$$
$$L_1C_1 = kD, \tag{A.213}$$
$$C_1(R_1 + R_2) = kE, \tag{A.214}$$
$$1 = kF, \tag{A.215}$$

where k is a positive constant. It can be calculated that

$$K = k^{-4}\, L_1C_1(C_1R_1R_2 - L_1)^2, \tag{A.216}$$
$$\lambda_1 = -k^{-3}\, C_1R_2[C_1R_2(R_1 + R_2) - 2L_1], \tag{A.217}$$
$$\delta = k^{-2}\, C_1[C_1(R_1 + R_2)(R_1R_2 + R_2R_3 + R_1R_3) - L_1(R_1 + R_2 + 4R_3)], \tag{A.218}$$
$$\tau_1 = -k^{-4}\, C1^2L_1R_3[C_1(R_1 + R_2)(2R_1R_2 + R_1R_3 + R_2R_3) - 2L_1(R_1 + R_2 + 2R_3)], \tag{A.219}$$
$$\gamma = -k^{-4}\, L_1C_1^2[C_1(R_1 + R_2)^2 - 4L_1], \tag{A.220}$$
$$\zeta_1 = k^{-3}\, C_1[C_1R_2^2(R_1 + R_2) + L_1(R_1 - 3R_2)], \tag{A.221}$$
$$AF - CD = k^{-2}\, L_1C_1(R_1 - R_2), \tag{A.222}$$

and hence $K \geq 0$ is necessary. From (A.215) we obtain

$$k = \frac{1}{F}. \tag{A.223}$$

Substituting (A.223) into (A.214), we obtain (A.208), which can be used in (A.213) to obtain (A.207). Eliminating L_1C_1 from (A.210) and (A.213) we obtain (A.205), and substituting (A.223) into (A.212) we get (A.206). We now have expressions for network elements R_1, R_2, L_1, C_1, and k which only contain R_3, together with A, B, C, D, E, and F. Substituting such expressions into (A.211) we obtain the quadratic equation (A.203). The discriminant of (A.203), $\Delta = 4DE^2FK$, is always greater than or equal to zero, and the equation therefore has two real solutions, which are coincident if $K = 0$.

In considering the roots of (A.203) we subdivide according to whether $\gamma\tau_1 > 0$, $\gamma\tau_1 < 0$, or $\gamma\tau_1 = 0$. We further subdivide into the following seven cases in which there is at least one positive root of (A.203):

$$\begin{array}{ll}
\text{(i)} & \gamma > 0,\ \ \tau_1 > 0,\ \ \delta < 0\,,\\
\text{(ii)} & \gamma < 0,\ \ \tau_1 < 0,\ \ \delta > 0\,,\\
\text{(iii)} & \gamma > 0,\ \ \tau_1 < 0\,,\\
\text{(iv)} & \gamma < 0,\ \ \tau_1 > 0\,,\\
\text{(v)} & \gamma = 0,\ \ \tau_1\delta < 0\,,\\
\text{(vi)} & \tau_1 = 0,\ \ \gamma\,\delta < 0\,,\\
\text{(vii)} & \gamma = 0,\ \ \delta = 0,\ \ \tau_1 = 0\,.
\end{array}$$

The first two cases correspond to two positive roots of (A.203), (iii)–(vi) to one positive root, and (vii) to infinitely many. We will now show that cases (i)–(vii) are equivalent to (A.194)–(A.202).

Step 1
We will first show that $\gamma > 0$ implies $\tau_1 > 0$, and hence case (iii) cannot occur. From (A.220), $\gamma > 0$ is equivalent to $L_1 > C_1(R_1+R_2)^2/4$, and from (A.219), $\tau_1 > 0$ is equivalent to $2L_1(R_1+R_2+2R_3) - C_1(R_1+R_2)(2R_1R_2+R_1R_3+R_2R_3) > 0$. Replacing L_1 in the latter expression by its strict lower bound (implied by $\gamma > 0$) gives

$$\begin{aligned}
&C_1(R_1+R_2)^2(R_1+R_2+2R_3)/2 - C_1(R_1+R_2)(2R_1R_2+R_1R_3+R_2R_3)\\
&= C_1(R_1-R_2)^2(R_1+R_2)/2 \geq 0\,.
\end{aligned}$$

Hence $\gamma > 0$ implies $\tau_1 > 0$.

Step 2
We now show that $\gamma \geq 0$, $\tau_1 > 0$ implies $\delta < 0$, and hence the latter condition can be removed from (i) and from the subcase in which $\tau_1 > 0$ in (v). It can be verified that the following identity holds:

$$(AF+CD)DF\,\delta = -DF(AF-CD)^2 - DF\tau_1 - AC\gamma\,, \tag{A.224}$$

from which the result follows.

Step 3
We will now show that $\tau_1 < 0$, $\delta > 0$ is impossible when $\gamma = 0$, and hence case (v) can be reduced to $\gamma = 0$, $\tau_1 > 0$ (having already shown in Step 2 that the condition $\delta < 0$ may be omitted). From (A.220), $\gamma = 0$ if $L_1 = C_1(R_1+R_2)^2/4$. Substituting this expression for L_1 into (A.218) and (A.219) yields

$$\tau_1 = k^{-4}\,\frac{C_1^4(R_1+R_2)^3(R_1-R_2)^2R_3}{8} = -2DF\,R_3\,\delta\,; \tag{A.225}$$

hence $\tau_1 \geq 0$ and $\delta \leq 0$, and the result follows. It is then easily seen that cases (i), (iv), and (v) taken together correspond to the single condition $\tau_1 > 0$.

Step 4
We now show that $\tau_1 > 0$ implies $\lambda_1 > 0$ if $AF - CD \geq 0$ (respectively, $\lambda_3 > 0$ if $AF - CD \leq 0$). From (A.219), $\tau_1 > 0$ is equivalent to

$$L_1 > \frac{C_1(R_1+R_2)(2R_1R_2+R_1R_3+R_2R_3)}{2(R_1+R_2+2R_3)}$$

and from (A.217), $\lambda_1 > 0$ is equivalent to $2L_1 - C_1R_2(R_1 + R_2) > 0$. Replacing L_1 in the latter expression by its strict lower bound (implied by $\tau_1 > 0$) gives

$$\begin{aligned}&\frac{C_1(R_1 + R_2)(2R_1R_2 + R_1R_3 + R_2R_3)}{R_1 + R_2 + 2R_3} - C_1R_2(R_1 + R_2)\\&= \frac{C_1(R_1^2 - R_2^2)(R_2 + R_3)}{R_1 + R_2 + 2R_3},\end{aligned}$$

which is greater than or equal to zero if $AF - CD \geq 0$ (see (A.222)).

Hence, we can conclude that cases (i), (iv), and (v) are equivalent to the subcase of (A.197) in which $\tau_1 > 0$, $\lambda_1 > 0$ if $AF - CD \geq 0$ (or, if $AF - CD \leq 0$, to the subcase of (A.202) in which $\tau_1 > 0$, $\lambda_3 > 0$).

Step 5

We will now show that $\tau_1 = 0$ implies $\delta \geq 0$ and $\gamma \leq 0$. From (A.219), if $\tau_1 = 0$, then

$$L_1 = \frac{C_1(R_1 + R_2)(2R_1R_2 + R_1R_3 + R_2R_3)}{R_1 + R_2 + 2R_3}.$$

Substituting this value for L_1 into (A.218) and (A.220) yields

$$\delta = k^{-2}\,\frac{C_1^2(R_1 + R_2)(R_1 - R_2)^2R_3}{2(R_1 + R_2 + 2R_3)} = \frac{-R_3}{2DF}\gamma$$

and the result follows. From (A.224) it also follows that $\tau_1 = 0$ and $\delta > 0$ always imply $\gamma < 0$, and hence the latter condition can be omitted in this case. Therefore, case (vi) reduces to $\tau_1 = 0$, $\delta > 0$.

Step 6

It can be shown that condition $\tau_1 = 0$ also implies $\lambda_1 > 0$ if $AF - CD > 0$, and $\lambda_3 > 0$ if $AF - CD < 0$. This is clear from the proof in Step 4 if we consider $\tau_1 \geq 0$ and $AF - CD \neq 0$. The case $AF - CD = 0$ (which corresponds to $R_1 = R_2$) cannot happen when $\tau_1 = 0$ and $\delta > 0$, since

$$\tau_1 = -4k^{-4}\,C_1^2L_1R_3(R_1 + R_3)(C_1R_1^2 - L_1), \tag{A.226}$$

$$\delta = 2k^{-2}\,(R_1 + 2R_3)(C_1R_1^2 - L_1) \tag{A.227}$$

when $R_1 = R_2$. Hence, case (vi) is equivalent to (A.195) and (A.200).

Step 7

We now turn to case (ii) and prove that $\zeta_1 > 0$ is necessary if $AF - CD \geq 0$ (respectively, $\zeta_3 > 0$ is necessary if $AF - CD \leq 0$). From (A.220), condition $\gamma < 0$ is equivalent to $C_1 > 4L_1/(R_1 + R_2)^2$ and from (A.221), $\zeta_1 > 0$ is equivalent to $C_1R_2^2(R_1 + R_2) + L_1(R_1 - 3R_2) > 0$. Replacing C_1 in the latter expression by its strict lower bound (implied by $\gamma < 0$) gives

$$4L_1R_2^2/(R_1 + R_2) + L_1(R_1 - 3R_2) = L_1(R_1 - R_2)^2/(R_1 + R_2) \geq 0,$$

and hence $\zeta_1 > 0$.

Step 8

We now show that conditions $\delta > 0$ and $\zeta_1 > 0$ always imply $\gamma < 0$ if $AF - CD \geq 0$

(the same result holds if $\delta > 0$, $\zeta_3 > 0$ and $AF - CD \leq 0$). Condition $\delta > 0$ is equivalent to $BE - 2(AF + CD) > 0$, while $\zeta_1 > 0$ can be written as $-BEF + CE^2 + 2F(AF - CD) > 0$. Using these two inequalities we get

$$C(E^2 - 4DF) > BEF - 2F(AF + CD) > 0\,,$$

and hence $\gamma = -DF(E^2 - 4DF) < 0$. Therefore, combining with Step 7, we see that the condition on γ may be replaced by $\zeta_1 > 0$ (respectively, $\zeta_3 > 0$) in case (ii).

Step 9

We now consider case (ii) with the condition $AF - CD \geq 0$ and subdivide it into three cases: $\lambda_1 > 0$, $\lambda_1 < 0$, and $\lambda_1 = 0$ (the case $AF - CD \leq 0$ is analogous with λ_1, ζ_1 replaced by λ_3, ζ_3). In this step we consider the case $\lambda_1 > 0$. It can be proven that, in this case, $AF - CD > 0$ necessarily as follows. If $AF - CD = 0$ (i.e., $R_1 = R_2$), then τ_1 is given by (A.226) and from (A.217) it follows that

$$\lambda_1 = -2k^{-3}\, C_1 R_1 (C_1 R_1^2 - L_1)\,. \tag{A.228}$$

Therefore, λ_1 and τ_1 necessarily have the same sign when $AF - CD = 0$, which is a contradiction in this case. Hence, if $\lambda_1 > 0$, case (ii) is equivalent to (A.196) (or if $\lambda_3 > 0$ when $AF - CD \leq 0$, case (ii) is equivalent to (A.201)). We will need to retain the condition $K \geq 0$ in these cases, since it is not implied by other conditions.

Step 10

If $\lambda_1 < 0$ in case (ii), we now show that $\delta > 0$ holds automatically (so it can be omitted) when $AF - CD \geq 0$ (the same is true if $\lambda_3 < 0$ when $AF - CD \leq 0$). Condition $\tau_1 < 0$ can be written as

$$BE(AF + CD) > ACE^2 + (AF + CD)^2\,,$$

while $\lambda_1 < 0$ is equivalent to $A\lambda_1 + BE(AF + CD) - BE(AF + CD) < 0$, which can be written as

$$BE(AF + CD) < ACE^2 + BCDE + AF(AF - CD)\,.$$

Comparing the lower and upper bounds of $BE(AF + CD)$ above implies $CD(BE - 3AF - CD) > 0$. It follows that

$$\delta = BE - 2(AF + CD) > 3AF + CD - 2(AF + CD) = AF - CD \geq 0\,,$$

and hence $\delta > 0$. We also note, from the expressions for ζ_1, ζ_3 and λ_1, λ_3 in Table 6.10, that

$$\zeta_1 = -\lambda_1 + F(AF - CD)\,, \tag{A.229}$$

$$\zeta_3 = -\lambda_3 - D(AF - CD)\,; \tag{A.230}$$

hence $\lambda_1 < 0$ implies $\zeta_1 > 0$ and $\lambda_3 < 0$ implies $\zeta_3 > 0$, and hence the condition on ζ_1 (respectively, ζ_3) can be omitted if the condition $\lambda_1 < 0$ (respectively, $\lambda_3 < 0$) is included. Therefore, if $\lambda_1 < 0$ and $AF - CD \geq 0$, case (ii) is equivalent to the subcase of (A.197) in which $\tau_1 < 0$, $\lambda_1 < 0$ (the other subcase being covered by Step 4). Similarly, if $\lambda_3 < 0$ and $AF - CD \leq 0$, case (ii) is equivalent to the subcase of (A.202) in which $\tau_1 < 0$, $\lambda_3 < 0$.

Step 11

Finally, if $\lambda_1 = 0$ in case (ii) with $AF - CD \geq 0$, we observe that $AF - CD > 0$ necessarily. This follows from Step 9, where it was shown that τ_1 and λ_1 have the same sign when $AF - CD = 0$, and hence $\tau_1 < 0$, $\lambda_1 = 0$, $AF - CD = 0$ cannot occur in this case.

It now follows as in Step 10 that $\delta > 0$ holds automatically, with the relaxed condition $\lambda_1 \leq 0$, and so it can be omitted in this case. Therefore, as in Step 10, from the identities (A.229) and (A.230) it is clear that if $\lambda_1 = 0$, then $\zeta_1 > 0$ (and if $\lambda_3 = 0$, then $\zeta_3 > 0$), and hence the condition on ζ_1 (respectively, ζ_3) can be omitted if the condition $\lambda_1 = 0$ (respectively, $\lambda_3 = 0$) is included.

In summary, if $\lambda_1 = 0$ when $AF - CD \geq 0$, case (ii) is equivalent to (A.194) (or if $\lambda_3 = 0$ when $AF - CD \leq 0$, case (ii) is equivalent to (A.199)).

Step 12

We show next that case (vii) is equivalent to (A.198), i.e.,

$$\gamma = 0,\;\; \delta = 0,\;\; \tau_1 = 0 \quad \Leftrightarrow \quad AF - CD = 0,\;\; K = 0\,.$$

If $AF - CD = 0$ (i.e., $R_1 = R_2$ from (A.222)), then τ_1 and δ have the expressions shown in (A.226) and (A.227), respectively, and

$$\gamma = -4k^{-4}\, C_1^2 L_1 (C_1 R_1^2 - L_1)\,. \tag{A.231}$$

It follows that if $K = 0$ (i.e., $L_1 = C_1 R_1 R_2$ from (A.216)), then $\gamma = 0$, $\delta = 0$, and $\tau_1 = 0$. We now show that $\gamma = 0$, $\delta = 0$, $\tau_1 = 0 \Rightarrow AF - CD = 0$, $K = 0$. From (A.220), $\gamma = 0$ if and only if

$$L_1 = C_1 (R_1 + R_2)^2 / 4\,. \tag{A.232}$$

It was shown in Step 3 that using this value for L_1 we obtain the expressions for τ_1 and δ in (A.225), from which we see that $R_1 = R_2$. It follows from (A.222) that $AF - CD = 0$, and from (A.232) that $L_1 = C_1 R_1^2$, and hence $K = 0$ (from (A.216)).

Step 13

We finally turn to the necessity condition $K \geq 0$ and show that it can be omitted in the conditions of the theorem (i.e., it is implied by the other conditions) in all but two cases, namely (A.196) and (A.201), other than case (A.198), which is already dealt with in Step 12.

If $AF - CD \geq 0$, we first show that $\lambda_1 \leq 0$ implies $K \geq 0$ (the same result holds if $AF - CD \leq 0$ and $\lambda_3 \leq 0$). It can be verified from the expressions in Table 6.10 that

$$FK = -\lambda_1 (AF - CD) + D(BF - CE)^2\,,$$

from which the result follows. Therefore, $K \geq 0$ can be omitted from (A.194) and from the subcase of (A.197) in which $\lambda_1 < 0$ (respectively, from (A.199) and from the subcase of (A.202) in which $\lambda_3 < 0$).

We next show that $\tau_1 \geq 0$ implies $K \geq 0$. From the expression in Table 6.10, $\tau_1 \geq 0$ implies $K \geq DF(B^2 - 4AC)$. If $B^2 - 4AC \geq 0$, we can immediately conclude that $K \geq 0$. Otherwise, if $B^2 - 4AC < 0$, it follows that

$$\begin{aligned} 4ACK &= 4AC[(AF - CD)^2 - (AE - BD)(BF - CE)] \\ &> B^2 (AF - CD)^2 - 4AC(AE - BD)(BF - CE) \\ &= (ABF - 2ACE + BCD)^2 \geq 0\,. \end{aligned}$$

Therefore, $K \geq 0$ can be omitted from (A.195), (A.200) and from the remaining subcases in (A.197), (A.202).

Sufficiency. Given a positive-real impedance (3.1) with $A, B, C, D, E, F > 0$, we can calculate one or two sets of solutions (depending on the signs of τ_1 and γ) for R_1, R_2, R_3, L_1, C_1, and k in cases (A.194)–(A.197) and (A.199)–(A.202), using (A.203), (A.205)–(A.208), and (A.223), and an infinite number of solutions in case (A.198). It remains to show that the necessary conditions derived are sufficient to ensure positivity of R_1 and R_2, and hence L_1 and C_1 from (A.207) and (A.208).

From (A.205), R_1 is positive if and only if $R_3 < A/D$, and, from (A.206), R_2 is positive if and only if $R_3 < C/F$. Therefore, in order for both R_1 and R_2 to be positive, we need

$$R_3 < \frac{C}{F} \quad \text{if } AF - CD \geq 0\,, \tag{A.233}$$

$$R_3 < \frac{A}{D} \quad \text{if } AF - CD \leq 0\,. \tag{A.234}$$

We consider the case that $AF - CD \geq 0$ (if $AF - CD \leq 0$, the proof is analogous, with λ_1 replaced by λ_3 and ζ_1 by ζ_3). We denote the two roots of (A.203) as

$$R_{3_A} = \frac{-2DF\,\delta - \sqrt{\Delta}}{2\gamma}\,, \qquad R_{3_B} = \frac{-2DF\,\delta + \sqrt{\Delta}}{2\gamma} \tag{A.235}$$

when $\gamma \neq 0$.

If conditions (A.194)–(A.196) or (A.197) hold (with $\tau_1 < 0$ and $\lambda_1 < 0$ in the latter case), then $\gamma < 0$ (see Steps 5 and 8) and $R_{3_A} \geq R_{3_B}$. We note that in case (A.195), $R_{3_B} = 0$, while in the other cases both R_{3_A} and R_{3_B} are strictly positive. Inequality (A.233) reduces to

$$\sqrt{\Delta} < 2D\zeta_1 \qquad (R_3 = R_{3_A})\,, \tag{A.236}$$

$$\sqrt{\Delta} > -2D\zeta_1 \qquad (R_3 = R_{3_B})\,. \tag{A.237}$$

Both sides of (A.236) are positive, since $\zeta_1 > 0$ if $\gamma < 0$ (this was proven in Step 7), and the inequality can therefore be squared and reduced to

$$4F\,\gamma\,(AF - CD)\,\lambda_1 < 0\,, \tag{A.238}$$

which holds for cases (A.195) and (A.196) but not for (A.194) and the subcase of (A.197). Inequality (A.237) need not be verified for case (A.195) (since $R_{3_B} = 0$ in this case), and is immediately satisfied for the other cases, since $\zeta_1 > 0$. Therefore, in conclusion, both roots of (A.203) yield positive values for R_1 and R_2 in case (A.196), and only the smallest positive root in cases (A.194), (A.195), and (A.197) in the subcase with $\tau_1 < 0$ and $\lambda_1 < 0$.

We now turn to the subcase of (A.197) with $\tau_1 > 0$ and $\lambda_1 > 0$. In this case nothing can be concluded on the sign of γ, and so we examine three cases on the sign of γ separately.

If $\gamma < 0$, then we see directly from (A.203) that R_{3_A} is the only positive root of (A.203), and inequality (A.233) reduces to (A.236). We will now show that in this case $\zeta_1 > 0$. Condition $\tau_1 > 0$ can be written as

$$-BE(AF + CD) > -ACE^2 - (AF - CD)^2 - 4ACDF\,.$$

It follows that

$$\begin{aligned}\zeta_1(AF+CD) &= -BEF(AF+CD)+CE^2(AF+CD)+2F(A^2F^2-C^2D^2)\\ &> -F(AF-CD)^2-4ACDF^2+C^2DE^2+2F(A^2F^2-C^2D^2)\,.\end{aligned}$$

Knowing that $\gamma < 0$ (i.e., $E^2 > 4DF$), if E^2 is replaced by $4DF$, the lower bound becomes

$$\begin{aligned}&-F(AF-CD)^2-4ACDF^2+4C^2D^2F+2F(A^2F^2-C^2D^2)\\ &= F(AF-CD)^2 \geq 0\,,\end{aligned}$$

and hence $\zeta_1 > 0$. Therefore, both sides of (A.236) are positive and the inequality can be squared and reduced to (A.238), which holds if $AF-CD \neq 0$. If $AF-CD=0$, it can be verified that the following identity holds:

$$C^2\,\gamma - F^2\tau_1 = F(AF+CD)\,\lambda_1\,. \tag{A.239}$$

From (A.239) it follows that the case $AF-CD=0$, $\gamma<0$, $\tau_1>0$, $\lambda_1>0$ cannot occur.

If $\gamma > 0$, the quadratic (A.203) has two positive roots $R_{3_B} \geq R_{3_A}$, since $\delta < 0$ (see Step 2), and inequality (A.233) reduces to

$$\sqrt{\Delta} > 2D\zeta_1 \qquad (R_3 = R_{3_A})\,, \tag{A.240}$$

$$\sqrt{\Delta} < -2D\zeta_1 \qquad (R_3 = R_{3_B})\,. \tag{A.241}$$

If $\zeta_1 \geq 0$, inequality (A.241) does not hold, and if $\zeta_1 < 0$, the same inequality can be squared and reduced to (A.238), which again does not hold. We now turn to (A.240). If $\zeta_1 < 0$, inequality (A.240) holds, and if $\zeta_1 \geq 0$, the inequality can be squared and reduced to

$$4F\,\gamma\,(AF-CD)\,\lambda_1 > 0\,,$$

which again holds, providing $AF-CD \neq 0$. We therefore need to check the case $AF-CD=0$ directly. If $AF-CD=0$, then $\zeta_1 = -\lambda_1$ from the expression in Table 6.10. Therefore, if the conditions of case (A.197) hold with $\tau_1>0$, $\lambda_1>0$, $\gamma>0$, and $AF-CD=0$, then $\zeta_1<0$, and inequality (A.240) is immediately satisfied.

If $\gamma=0$, the only root of (A.203) is $R_3 = -\tau_1/(2DF\,\delta) > 0$, since $\delta<0$ (see Step 2), and inequality (A.233) reduces to $\tau_1 + 2CD\,\delta < 0$. It can be verified that

$$\tau_1 + 2CD\,\delta = -C^2\,\gamma/F^2 - \lambda_1\,(AF-CD)/F\,,$$

and hence, for $\gamma=0$, the inequality reduces to

$$\tau_1 + 2CD\,\delta = -\lambda_1(AF-CD)/F < 0\,,$$

which holds if $AF-CD \neq 0$, since $\lambda_1 > 0$. It can also be verified that the following identity always holds for $AF-CD=0$ and $\gamma=0$:

$$\tau_1 = -A\,\lambda_1 - CE^2\,K/\lambda_1\,. \tag{A.242}$$

From (A.242) it follows that the case $AF - CD = 0$, $\tau_1 > 0$, $\lambda_1 > 0$ cannot occur. We can therefore conclude from the analysis of the three cases $\gamma < 0$, $\gamma > 0$, and $\gamma = 0$ above that if the conditions of case (A.197) hold with $\tau_1 > 0$ and $\lambda_1 > 0$, then solution R_{3_A} yields positive values for R_1 and R_2 if $\gamma \neq 0$, while if $\gamma = 0$, the only root of (A.203) yields positive values for R_1 and R_2. This establishes that R_3 should be the smallest positive root in case (A.197).

In case (A.198) we have $\gamma = 0$, $\delta = 0$, and $\tau_1 = 0$ (see Step 12), and hence any value of x satisfies (A.203). Since $A/D = C/F$ in this case, inequalities (A.233) and (A.234) are equivalent, and the theorem statement follows (i.e., R_3 can be chosen arbitrarily provided that $R_3 < C/F$). □

Figure A.10 shows the realizability region for the network, plotted on the (U, V)-plane for $W = 0.8$ (i.e., $AF - CD > 0$). The expressions in terms of U, V, W for all the symbols appearing in the figure can be found in Table 6.10. It is clear from the figure that the curve γ_c does not act as a boundary for the realizability region, and it was in fact possible in the proof to eliminate γ from the realizability conditions. Curves δ_c and ζ_c are not active boundaries either but are still needed to properly define the region corresponding to cases (A.195) and (A.196). We note that it may be possible to write the conditions in different ways or to further simplify such conditions.

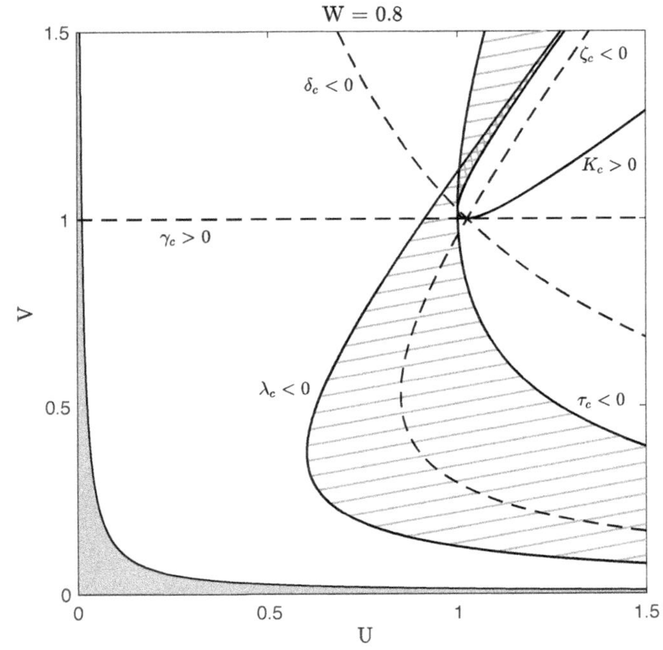

Figure A.10. *Realizability region for the network on the (U, V)-plane for $W = 0.8$. The hatched regions correspond to case (A.197), while the crossed region corresponds to case (A.196). Cases (A.194) and (A.195) correspond to the boundaries of the crossed region, with $\lambda_c = 0$ and $\tau_c = 0$, respectively. The dashed curves, namely γ_c, δ_c, and ζ_c, are not active boundaries for the realizability region.*

A.9 ▪ Equivalence class V_H^1

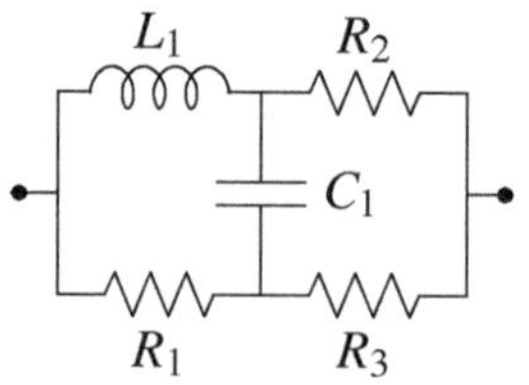

Figure A.11. *Network #70 from subfamily V_H. By Theorem* 3 *in* [47] *it can only realize regular impedances.*

Theorem A.9. *The positive-real biquadratic impedance* (3.1) *with A, B, C, D, E, $F > 0$ can be realized as in Figure* A.11, *with R_1, R_2, R_3, L_1, and C_1 positive and finite, if and only if*

$$AF - CD > 0\,, \tag{A.243}$$

$$\mu_1 \geq 0\,, \tag{A.244}$$

and

$$\textit{signs of } \lambda_1, \lambda_2, \rho_1 \textit{ not all the same}\,, \tag{A.245}$$

where λ_1, λ_2, μ_1, and ρ_1 are defined in Table 6.10. *When one of λ_1, λ_2, ρ_1 equals zero, the other two must have different algebraic signs or both equal zero, i.e.,*

$$\lambda_1 = \lambda_2 = 0, \qquad \rho_1 = 0\,. \tag{A.246}$$

If conditions (A.243)–(A.245) *are satisfied, then R_3 is any positive root of the quadratic equation in x*

$$D\lambda_1 x^2 + \rho_1 x + C\lambda_2 = 0\,, \tag{A.247}$$

while if conditions (A.243), (A.244), *and* (A.246) *are satisfied, R_3 is any positive value. R_2 is the positive root of the quadratic equation in y*

$$c_1\, y^2 + c_2\, y + c_3 = 0\,, \tag{A.248}$$

where

$$\begin{aligned} c_1 &= AF - CD\,, \\ c_2 &= (A + DR_3)(FR_3 - C)\,, \\ c_3 &= -CR_3(A + DR_3)\,, \end{aligned}$$

and

$$R_1 = \frac{R_2(A - DR_3) + AR_3}{D(R_2 + R_3)}\,, \tag{A.249}$$

$$L_1 = \frac{DR_1(R_1R_2 + R_1R_3 + R_3^2)}{(R_2 + R_3)\big[E(R_1 + R_3) - B\big]}\,, \tag{A.250}$$

$$C_1 = \frac{D(R_1 + R_2 + R_3)}{F(R_2 + R_3)L_1}\,. \tag{A.251}$$

Proof. Note that this result was also proven in [47, Appendix B], where μ_1 and ρ_1 are called η_1 and η_2, respectively. We provide here an independent proof.

Necessity. The impedance of the network shown in Figure A.11 is a biquadratic, which can be computed as

$$Z(s) = \frac{n(s)}{d(s)}, \tag{A.252}$$

where

$$\begin{aligned} n(s) &= L_1C_1(R_1R_2 + R_1R_3 + R_2R_3)s^2 + \big(C_1R_1R_2R_3 + L_1(R_1 + R_3)\big)s \\ &\quad + R_2(R_1 + R_3), \\ d(s) &= L_1C_1(R_2 + R_3)s^2 + \big(C_1R_1(R_2 + R_3) + L_1\big)s + R_1 + R_2 + R_3. \end{aligned}$$

Equating impedance (A.252) with (3.1) we obtain, for a positive constant k,

$$L_1C_1(R_1R_2 + R_1R_3 + R_2R_3) = kA, \tag{A.253}$$
$$C_1R_1R_2R_3 + L_1(R_1 + R_3) = kB, \tag{A.254}$$
$$R_2(R_1 + R_3) = kC, \tag{A.255}$$
$$L_1C_1(R_2 + R_3) = kD, \tag{A.256}$$
$$C_1R_1(R_2 + R_3) + L_1 = kE, \tag{A.257}$$
$$R_1 + R_2 + R_3 = kF, \tag{A.258}$$

which are a set of necessary and sufficient conditions for (3.1) to be realized as in Figure A.11. It can be calculated that

$$AF - CD = k^{-2}\, R_1L_1C_1(R_1R_2 + R_1R_3 + 2R_2R_3 + R_3^2) > 0, \tag{A.259}$$
$$\begin{aligned} \mu_1 = k^{-4}\, L_1C_1\big[&R_1R_2C_1(R_1 + 2R_3)(R_2 + R_3) \\ &- L_1(R_1 + R_3)(2R_2 + R_3)\big]^2 \geq 0, \end{aligned} \tag{A.260}$$
$$K = k^{-4}\, L_1C_1\big[R_1^2R_2C_1(R_2 + R_3) + L_1R_3(R_1 + R_3)\big]^2 > 0, \tag{A.261}$$

and hence (A.243) and (A.244) are necessary.

From (A.258) we obtain

$$k = (R_1 + R_2 + R_3)/F, \tag{A.262}$$

and from (A.256) we get (A.251). Equations (A.254) and (A.257) now reduce to

$$\begin{aligned} F(R_1 + R_3)(R_2 + R_3)\, L_1^2 - B(R_1 + R_2 + R_3)(R_2 + R_3)\, L_1 \\ + DR_1R_2R_3(R_1 + R_2 + R_3) = 0, \\ FL_1^2 - E(R_1 + R_2 + R_3)\, L_1 + DR_1(R_1 + R_2 + R_3) = 0. \end{aligned}$$

Eliminating the term in FL_1^2 from the two equations and solving for L_1 gives (A.250). Eliminating L_1C_1 from (A.253) and (A.256) and solving for R_1 gives (A.249), and substituting (A.249) and (A.262) into (A.255) we get the quadratic equation (A.248), with $y = R_2$. From (A.248) an expression for R_2^2 can be found as follows:

$$R_2^2 = \frac{(A + DR_3)\big[(C - FR_3)R_2 + CR_3\big]}{AF - CD}. \tag{A.263}$$

We now have expressions for elements R_1, L_1, C_1 and for the constant k which only contain R_2 and R_3, together with A, B, C, D, E, and F, and an expression for R_2^2 which contains R_3, together with the same coefficients, and is linear in R_2. Substituting the expressions for R_1, L_1, C_1, and k into (A.257) we obtain a polynomial of fourth order in R_2 and fifth order in R_3. Then, substituting for R_2^2 from (A.263) twice to reduce the power of R_2, after further manipulation all terms in R_2 cancel out and the higher powers in R_3 are eliminated, and we obtain the quadratic equation (A.247) in R_3.

When λ_1 and λ_2 in (A.247) are of opposite sign, the quadratic has one real positive root. When λ_1 and λ_2 are of the same sign, the quadratic has two real positive roots if and only if ρ_1 is of opposite sign and the discriminant is nonnegative. It can be calculated that the discriminant of (A.247) is $K\mu_1$ and is therefore nonnegative by (A.260) and (A.261). It is easily seen that when one of the three coefficients in (A.247) is zero it is necessary that the other two have different sign, in order for R_3 to be positive. Similarly, when two of the three coefficients in (A.247) are zero it is necessary that the other one also equals zero, in order for R_3 to be nonzero. Therefore conditions (A.245) and (A.246) are necessary.

We finally show that the condition $K > 0$ is implied by (A.243) and (A.245)–(A.246) and can therefore be omitted from the theorem statement. It can be verified that the following identities always hold:

$$\begin{aligned} E^2 F\, K &= -\lambda_1 E^2 (AF - CD) + D\big(\lambda_1 + F(AF - CD)\big)^2, \\ AB^2\, K &= -\lambda_2 B^2 (AF - CD) + C\big(\lambda_2 + A(AF - CD)\big)^2; \end{aligned}$$

therefore when $\lambda_1 \leq 0$ or $\lambda_2 \leq 0$ it follows that $K > 0$. If $\lambda_1 > 0$ and $\lambda_2 > 0$, it follows from (A.245) that $\rho_1 < 0$, and hence $K = -\rho_1 + 2CD(AF - CD) > 0$.

Sufficiency. Given a positive-real impedance (3.1) with $A, B, C, D, E, F > 0$ satisfying (A.243)–(A.246), we can calculate from (A.247) one, two, or infinitely many positive solutions for R_3, depending on the signs of λ_1, λ_2, and ρ_1. We now show that any positive value of R_3 leads to positive values of the other network elements.

It is easily seen that coefficients c_1 and c_3 in (A.248) are of opposite sign; therefore the quadratic always has one real positive root $R_2 > 0$. From (A.249), R_1 is positive if

$$R_2(A - DR_3) > -AR_3\,. \tag{A.264}$$

If $R_3 \leq A/D$, then (A.264) is satisfied; otherwise it can be rewritten as

$$R_2 < \frac{-AR_3}{A - DR_3}. \tag{A.265}$$

Since R_2 is the only positive solution to the quadratic equation (A.248), which represents a parabola that opens up, inequality (A.265) holds if and only if (A.248) evaluated at $-AR_3/(A - DR_3)$ gives a positive value, that is,

$$c_1 \left(\frac{-AR_3}{A - DR_3}\right)^2 + c_2 \left(\frac{-AR_3}{A - DR_3}\right) + c_3 > 0\,. \tag{A.266}$$

After some manipulation, inequality (A.266) can be reduced to

$$\frac{R_3^4\,(AF - CD)\,D^2}{(A - DR_3)^2} > 0\,,$$

which always holds.

Hence R_1 is positive and, from (A.262), also k is positive. From (A.251), L_1 and C_1 have the same sign, and we can therefore conclude from (A.254) that they are both positive. □

Figure A.12 shows the realizability region for the network, plotted on the (U, V)-plane, for $W = 0.7$, while Figure A.13 shows the realizability region for $W = 0.25$. The expressions in terms of U, V, W for all the symbols appearing in the figures can be found in Table 6.10. It is clear from both figures that the curve ρ_c is not an active boundary but is still needed to properly define the realizability regions.

The interior of the hatched region is realizable with two distinct solutions, unless either λ_c, $\lambda_c^\dagger$, or ρ_c is zero, in which case there is only one solution. The boundary is realizable only for $\mu_c = 0$, where there are two coincident solutions. It can finally be verified that case (A.246), where there are infinite solutions, corresponds to the case $W = 1/3$. As pointed out in [47, 67], there exists a realizable region with $\lambda_c < 0$ and $\lambda_c^\dagger < 0$ (corresponding to nonregular impedances) only for $W \in (1/3, 1)$, as can be seen from Figures A.12 and A.13.

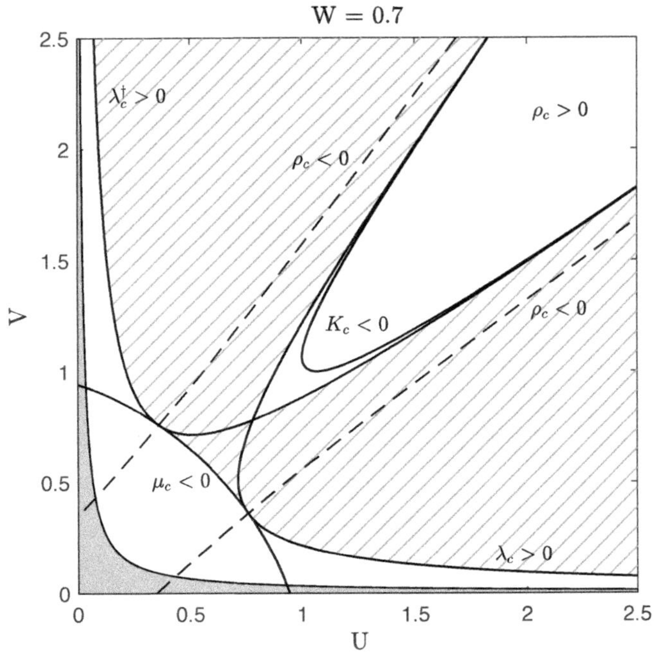

Figure A.12. *Realizability region for the network on the (U, V)-plane for $W = 0.7$. The region defined by $\lambda_c < 0$, $\lambda_c^\dagger < 0$, and $\mu_c \geq 0$ corresponds to nonregular impedances.*

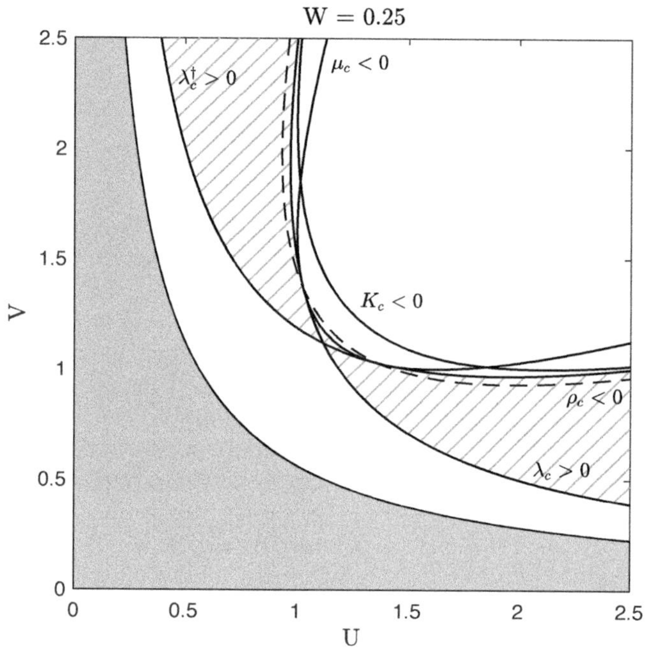

Figure A.13. *Realizability region for the network on the (U, V)-plane for $W = 0.25$.*

A.10 ▪ Equivalence class V_I

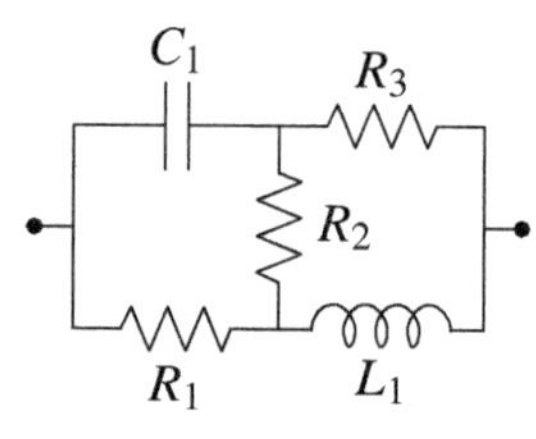

Figure A.14. *Network #108 from subfamily V_I. By Lemma 12 in [47] it can only realize regular impedances.*

Theorem A.10. *The positive-real biquadratic impedance (3.1) with A, B, C, D, E, $F > 0$ can be realized as in Figure A.14, with R_1, R_2, R_3, L_1, and C_1 positive and finite, if and only if*

$$K \geq 0\,, \tag{A.267}$$

where K is defined in (3.3), and the conditions of one of the following cases hold:

$$\textit{Case (1)} \quad \tau_1\tau_2 < 0\,, \tag{A.268}$$

$$\textit{Case (2)} \quad \tau_1 = 0,\ \ \tau_2 < 0,\ \ \psi > 0\,, \tag{A.269}$$

$$\textit{Case (3)} \quad \tau_2 = 0,\ \ \tau_1 < 0,\ \ \psi > 0\,, \tag{A.270}$$

$$\textit{Case (4)} \quad \tau_1 < 0,\ \ \tau_2 < 0,\ \ \psi > 0\,, \tag{A.271}$$

$$\textit{Case (5)} \quad \tau_1 = 0,\ \ \tau_2 = 0,\ \ \psi = 0\,, \tag{A.272}$$

where τ_1, τ_2, and ψ are defined in Table 6.10. *In cases* (A.268)–(A.270), *R_2 is the only positive root of the quadratic equation in x*

$$DF\,\tau_1\,x^2 + \psi\,x + AC\,\tau_2 = 0\,, \tag{A.273}$$

while if (A.271) *holds, then R_2 is either of the two positive roots of the same quadratic. If* (A.272) *holds, then R_2 is any positive value. The other elements of the network are given by*

$$R_1 = C/F\,, \tag{A.274}$$
$$R_3 = A/D\,, \tag{A.275}$$
$$L_1 = \frac{(DR_2 + A)(AF + CD)}{DEF\,R_2 + \gamma_2}\,, \tag{A.276}$$
$$C_1 = \frac{DR_2 + A}{L_1(FR_2 + C)}\,, \tag{A.277}$$

where $\gamma_2 = AEF - BDF + CDE$.

Proof. *Necessity.* The impedance of the network shown in Figure A.14 is a biquadratic, which can be computed as

$$Z(s) = \frac{n(s)}{d(s)}\,, \tag{A.278}$$

where

$$\begin{aligned} n(s) &= L_1C_1R_3(R_1 + R_2)s^2 + \big(C_1R_1R_2R_3 + L_1(R_1 + R_2 + R_3)\big)s \\ &\quad + R_1(R_2 + R_3)\,, \\ d(s) &= L_1C_1(R_1 + R_2)s^2 + \big(C_1(R_1R_2 + R_1R_3 + R_2R_3) + L_1\big)s + R_2 + R_3\,. \end{aligned}$$

Equating impedance (A.278) with (3.1), we obtain

$$L_1C_1R_3(R_1 + R_2) = kA\,, \tag{A.279}$$
$$C_1R_1R_2R_3 + L_1(R_1 + R_2 + R_3) = kB\,, \tag{A.280}$$
$$R_1(R_2 + R_3) = kC\,, \tag{A.281}$$
$$L_1C_1(R_1 + R_2) = kD\,, \tag{A.282}$$
$$C_1(R_1R_2 + R_1R_3 + R_2R_3) + L_1 = kE\,, \tag{A.283}$$
$$R_2 + R_3 = kF\,, \tag{A.284}$$

where k is a positive constant. From (A.284) we obtain

$$k = (R_2 + R_3)/F\,. \tag{A.285}$$

Eliminating $L_1C_1(R_1 + R_2)$ from (A.279) and (A.282) we obtain (A.275), while eliminating $R_2 + R_3$ from (A.281) and (A.284) we obtain (A.274). Solving (A.282) for C_1 then gives (A.277). Equations (A.280) and (A.283) now reduce to

$$\begin{aligned} (DFR_2 + AF + CD)(FR_2 + C)\,L_1^2 - B(DR_2 + A)(FR_2 + C)\,L_1 \\ + ACR_2(DR_2 + A) = 0\,, \\ DF(FR_2 + C)\,L_1^2 - E(DR_2 + A)(FR_2 + C)\,L_1 \\ + (DR_2 + A)\big((AF + CD)R_2 + AC\big) = 0\,. \end{aligned}$$

Eliminating the term in $(FR_2 + C)L_1^2$ from the two equations and solving for L_1 gives (A.276). We now have expressions for R_1, R_3, L_1, C_1, and k which only contain R_2, together with A, B, C, D, E, F. Substituting such expressions into (A.283) we obtain the quadratic equation (A.273) in R_2. It can be calculated that

$$K = k^{-4}\, L_1C_1(R_1 + R_2)^2(R_2 + R_3)^2(L_1 - R_1R_3C_1)^2\,,$$

from which (A.267) is necessary. The discriminant of (A.273),

$$\Delta = (AF + CD)^2\,(K + 4ACDF)\,K\,, \tag{A.286}$$

is therefore nonnegative from (A.267). If $\tau_1\tau_2 < 0$, then (A.273) has one real positive root (case (A.268)), while if $\tau_1\tau_2 > 0$, then there are two positive roots if and only if ψ is of the opposite sign to τ_1 and τ_2. It can be verified that the following identity always holds:

$$\psi = AF\,\tau_1 + CD\,\tau_2 + AD(BF - CE)^2\,. \tag{A.287}$$

Hence $\tau_1 > 0$, $\tau_2 > 0$ implies $\psi > 0$, from which it follows that ψ being of opposite sign to τ_1 and τ_2 corresponds only to (A.271). If $\tau_1 = 0$, then we need $\tau_2\psi < 0$ in order for R_2 to be positive. It follows from (A.287) that the case $\tau_2 > 0$, $\psi < 0$ cannot occur, and hence the case $\tau_1 = 0$, $\tau_2\psi < 0$ reduces to (A.269). A similar argument holds for case (A.270) when $\tau_2 = 0$. Finally, if $\tau_1 = 0$, $\tau_2 = 0$, $\psi = 0$ (i.e., case (A.272)), then any value of x solves (A.273), and so no restriction is placed on R_2.

Sufficiency. Given a positive-real impedance (3.1) with $A, B, C, D, E, F > 0$ satisfying (A.267), we can calculate from (A.273) one positive solution for R_2 in cases (A.268)–(A.270), two positive solutions in case (A.271), and infinitely many in case (A.272). We now show that in all cases the values obtained for R_2 lead to a positive L_1 and hence C_1 (from (A.277)).

From (A.276), L_1 will be positive if

$$R_2 > -\gamma_2/(DEF)\,, \tag{A.288}$$

and hence if $\gamma_2 \geq 0$, then (A.288) is immediately satisfied. It can be verified that the following identity always holds:

$$B\,\gamma_2 = (AF + CD)^2 - \tau_2\,, \tag{A.289}$$

and hence $\gamma_2 > 0$ (and therefore $L_1 > 0$) in all the cases in which $\tau_2 \leq 0$, i.e., in (A.269)–(A.272) and the subcase $\tau_1 > 0$, $\tau_2 < 0$ of (A.268). In the other subcase of (A.268), i.e., $\tau_1 < 0$, $\tau_2 > 0$, the only positive root of the quadratic (A.273) is

$$R_2 = \frac{-\psi - \sqrt{\Delta}}{2DF\,\tau_1}\,. \tag{A.290}$$

Substituting (A.290) into (A.288) we obtain the inequality

$$(AF + CD)\left[E(K + 4ACDF) - 2BDF(AF + CD)\right] < E\sqrt{\Delta}\,. \tag{A.291}$$

For (A.291) to hold it is sufficient that the inequality holds when both sides are squared, and after some manipulation the latter inequality reduces to

$$-4DF(BF - CE)(AE - BD)(AF + CD)^2\,\tau_1 < 0\,. \tag{A.292}$$

As mentioned above, if $\gamma_2 \geq 0$, inequality (A.288) is immediately satisfied. If $\gamma_2 < 0$, then $AE - BD < 0$ and $BF - CE > 0$ (directly from the expression for γ_2), and hence inequality (A.292) holds in this case. Therefore L_1 is positive for all sets of solutions. □

Figure A.15 shows the realizability region for the network on the (U, V)-plane for $W = 0.5$ (i.e., $AF - CD > 0$). The expressions in terms of U, V, W for all the symbols appearing in the figure can be found in Table 6.10. It is clear from the figure that the curve ψ_c is not an active boundary but is still needed to properly define the realizability region.

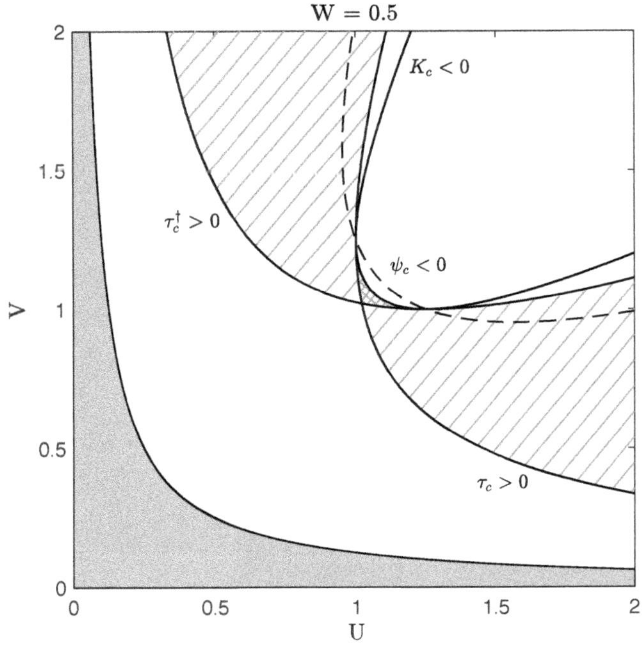

Figure A.15. *Realizability region for the network on the* (U, V)*-plane for* $W = 0.5$*. The hatched regions correspond to case* (A.268)*, while the crossed region corresponds to case* (A.271)*. Cases* (A.269) *and* (A.270) *correspond to the boundaries of the crossed region, with* $\tau_c = 0$ *and* $\tau_c^\dagger = 0$*, respectively. It can be verified that the conditions of case* (A.272) *imply* $AF - CD = 0$ *(i.e.,* $W = 1$*), and hence case* (A.272) *cannot be represented in this plot.*

Appendix B
Basic graphs

Below is the enumeration of all the basic graphs with at most five edges, with the corresponding network numbers from the Ladenheim canonical set. The superscript d indicates the graph dual, while values in brackets indicate the 40 networks which are eliminated from the canonical total (see Section 5.1).

Graph	One-element networks	R	L	C	Tot.	Canonical tot.	Network number
A		1	1	1	3	3	1, 2, 3
				Total	3	3	

Graph	Two-element networks	$2R$	R–L R–C	LC	Tot.	Canonical tot.	Network number
B		0	2	1	3	3	4, 5, 6
B^d		0	2	1	3	3	7, 8, 9
				Total	6	6	

Graph	Three-element networks	$3R$	$2R$–L $2R$–C	$2L$–R $2C$–R	LRC	Tot.	Canonical tot.	Network number
C		0	0	0	1	1	1	10
D	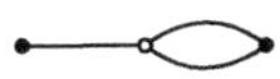	0	2	2	3	7	7	11, 14, 15, 17, 26, 27, 34
D^d	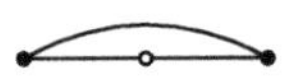	0	2	2	3	7	7	12, 13, 16, 18, 41, 42, 49
C^d		0	0	0	1	1	1	19
					Total	16	16	

Graph	Four-element networks	$3R$–L $3R$–C	$2R$–$2L$ $2R$–$2C$	$2R$–LC	Tot.	Canonical tot.	Network number
E		0	0	0	0	0	-
F		0	2	2	4	4	20, 25, 28, 32
G		(2)	4	5	11	9	21, 22, 24, 29, 30, 33, 63, 71, 87
H	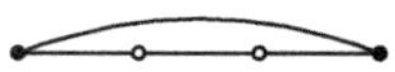	0	0	1	1	1	72
I	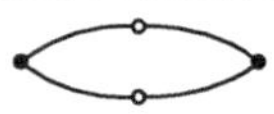	0	2	1	3	3	23, 31, 97
I^d		0	2	1	3	3	38, 46, 96
H^d		0	0	1	1	1	73
G^d		(2)	4	5	11	9	36, 37, 40, 44, 45, 48, 62, 74, 88
F^d	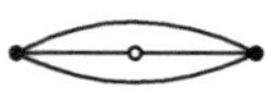	0	2	2	4	4	35, 39, 43, 47
E^d		0	0	0	0	0	-
				Total	38	34	

Graph	Five-element networks	$4R$–L $4R$–C	$3R$–$2L$ $3R$–$2C$	$3R$–LC	Tot.	Canonical tot.	Network number
J		0	0	0	0	0	-
K	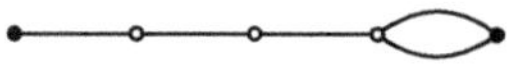	0	0	0	0	0	-
L	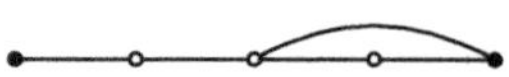	0	(2)	(2)	4	0	-
M		0	0	(1)	1	0	-
N	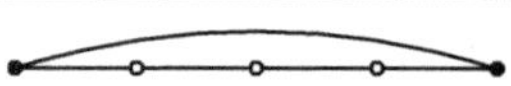	0	0	0	0	0	-
O	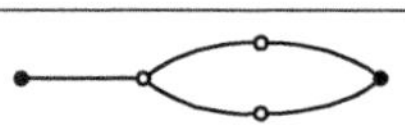	0	(2)	1	3	1	104
P	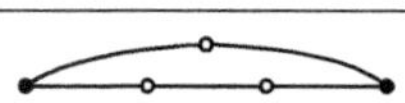	0	0	0	0	0	-

Graph	Five-element networks	$4R$–L $4R$–C	$3R$–$2L$ $3R$–$2C$	$3R$–LC	Tot.	Canonical tot.	Network number
Q		0	0	0	0	0	-
R		0	2	2	4	4	50, 68, 79, 93
S		(2)	(2) + 4	(3) + 4	15	8	51, 52, 64, 69, 80, 81, 89, 94
T		0	(2)	1	3	3	53, 82, 98
U		0	2	2	4	4	59, 83, 99, 100
V	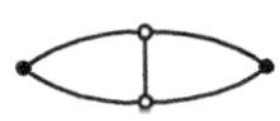	(4)	(4) + 4	5	17	9	60, 61, 70, 85, 86, 95, 105, 107, 108
U^d	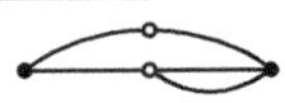	0	2	2	4	4	58, 84, 102, 103
T^d	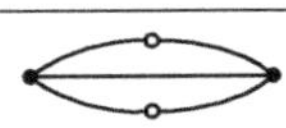	0	2	1	3	3	57, 78, 101
S^d		(2)	(2) + 4	(3) + 4	15	8	55, 56, 66, 67, 76, 77, 90, 92
R^d	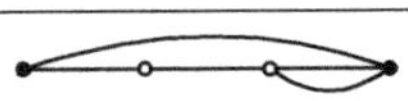	0	2	2	4	4	54, 65, 75, 91
Q^d		0	0	0	0	0	-
P^d		0	0	0	0	0	-
O^d	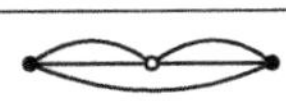	0	(2)	1	3	1	106
N^d	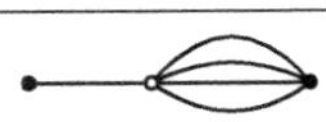	0	0	0	0	0	-
M^d	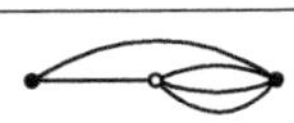	0	0	(1)	1	0	-
L^d	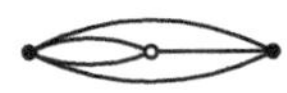	0	(2)	(2)	4	0	-
K^d	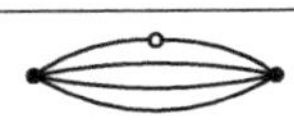	0	0	0	0	0	-
J^d	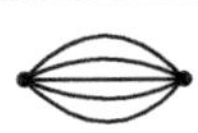	0	0	0	0	0	-
				Total:	85	49	
				Total:	148	108	

Appendix C

The Ladenheim networks (numerical order)

Network #1

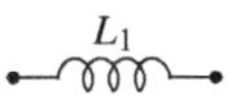

Network #2

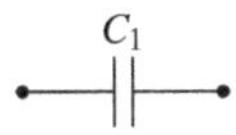

Network #3

Network #4

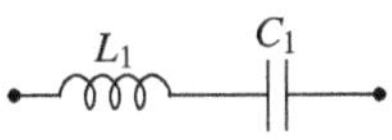

Network #5

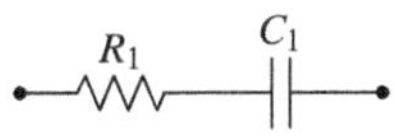

Network #6

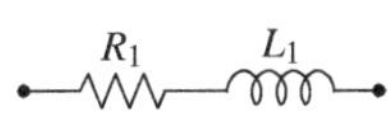

Network #7

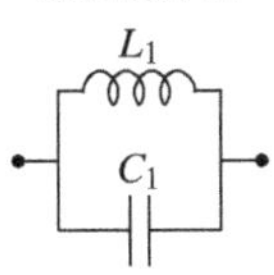

Network #8

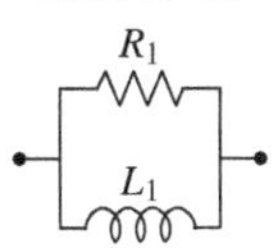

Network #9

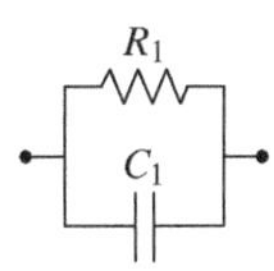

Network #10

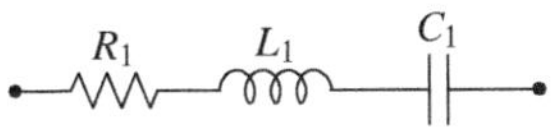

Network #11

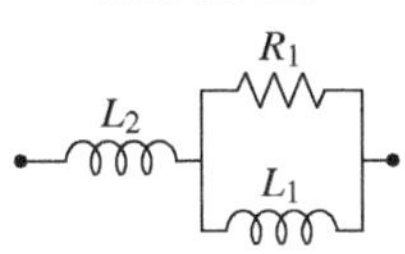

Network #12

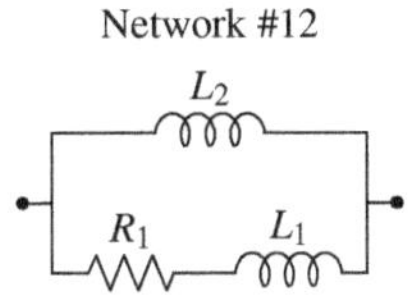

Network #13

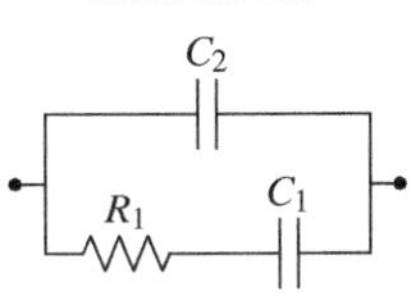

Network #14

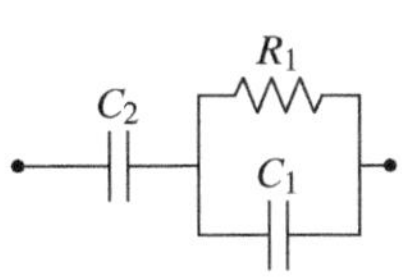

Network #15

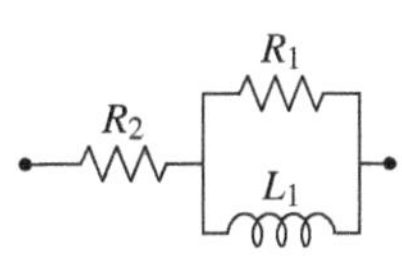

Network #16

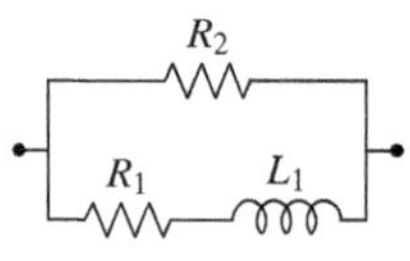

Network #17

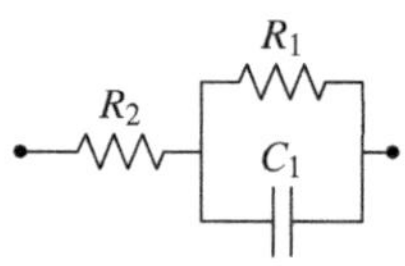

Network #18

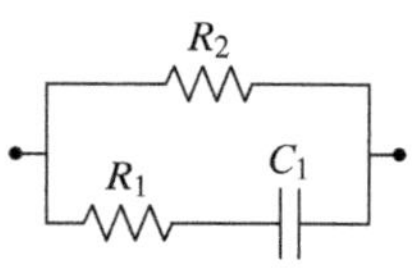

Network #19

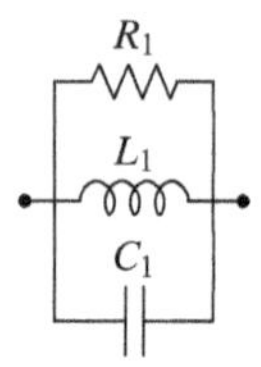

Network #20

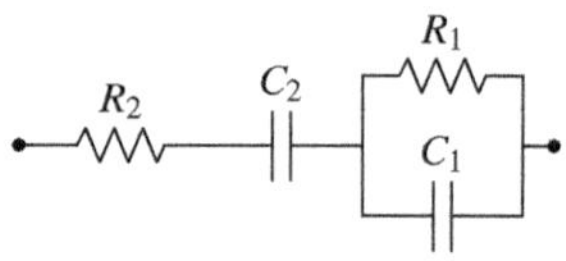

Network #21

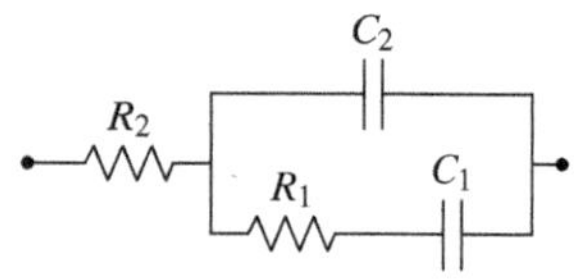

Network #22

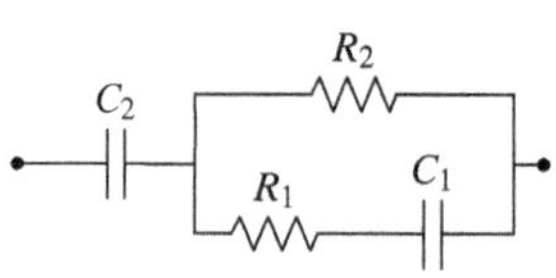

Network #23

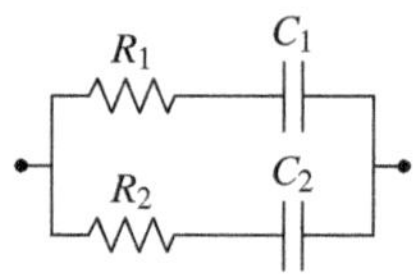

Network #24

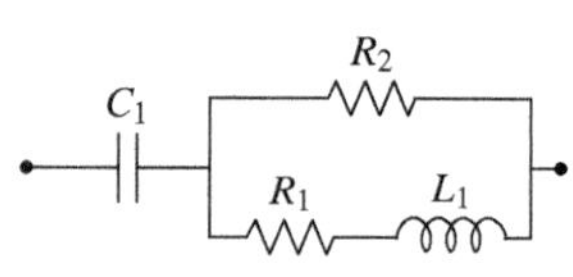

Network #25

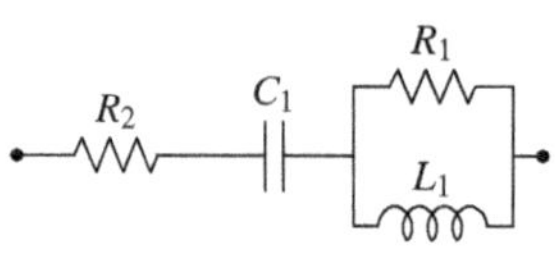

Network #26

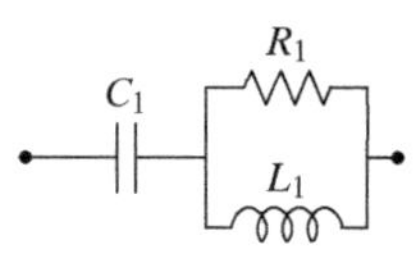

Network #27

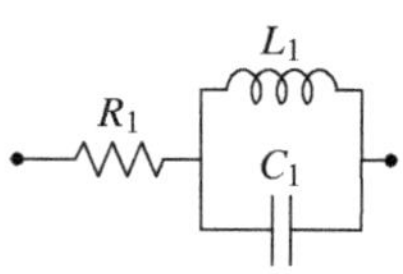

Network #28

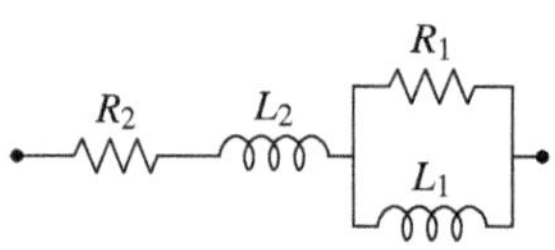

Network #29

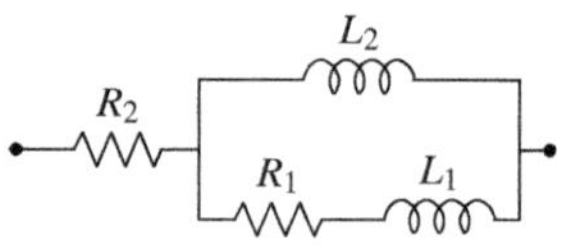

Network #30

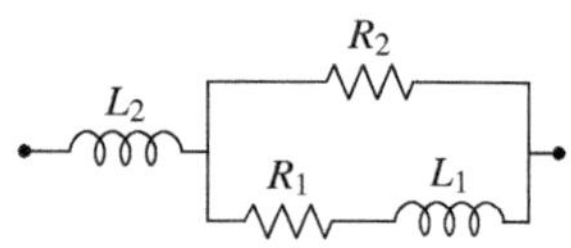

Network #31

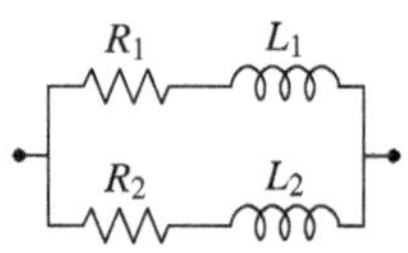

Network #32

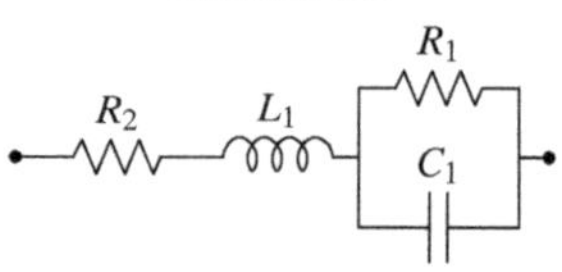

Network #33

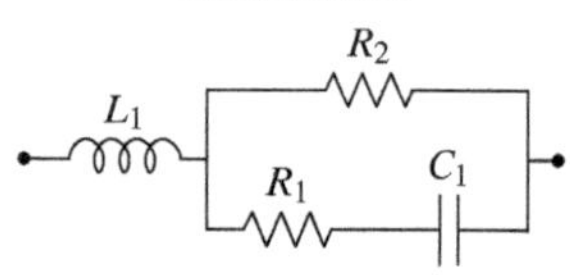

Network #34

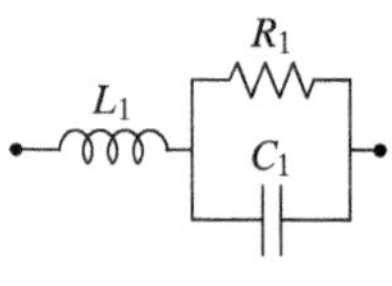

Network #35

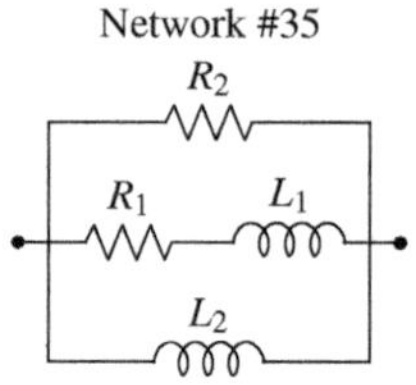

Network #36

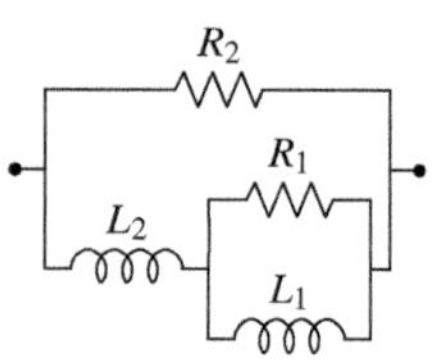

Network #37

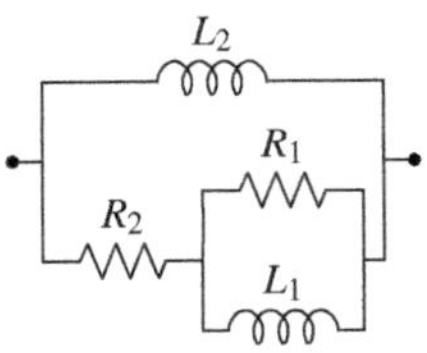

Network #38

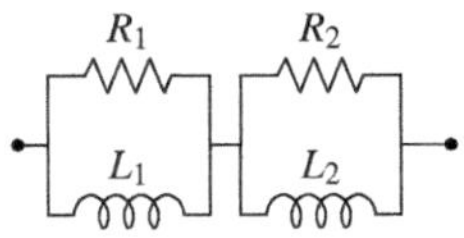

Network #39

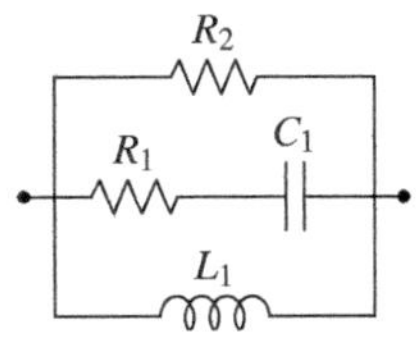

Network #40

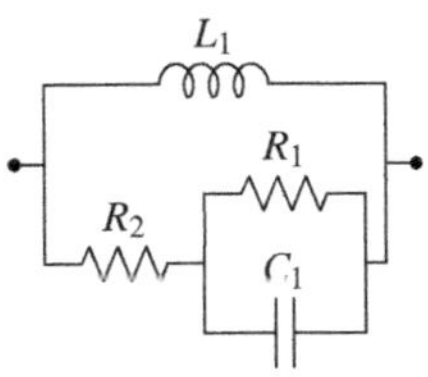

Network #41

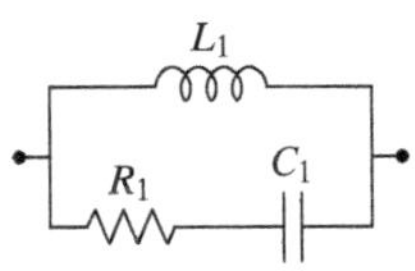

Network #42

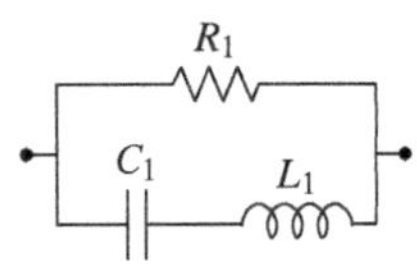

Network #43

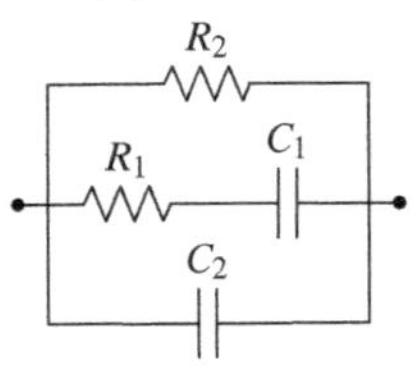

Network #44

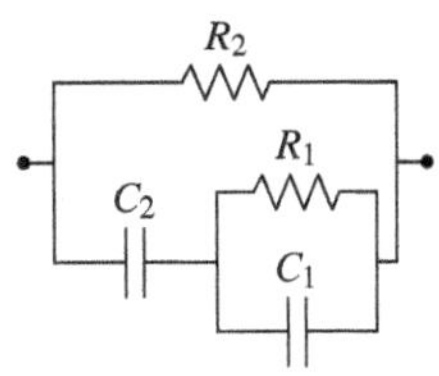

Network #45

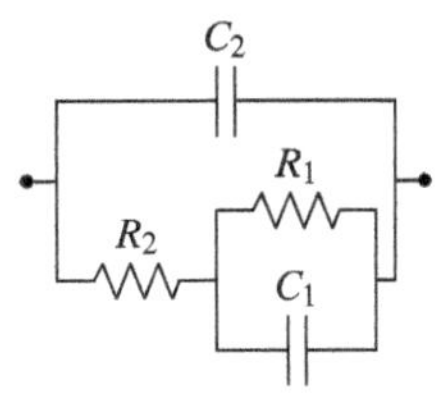

Network #46

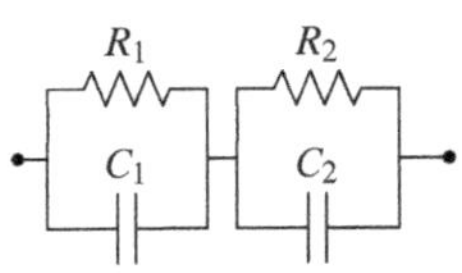

Network #47

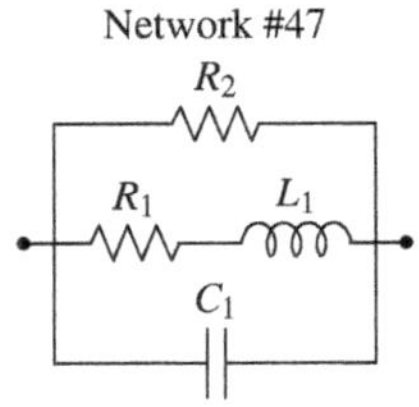

Network #48

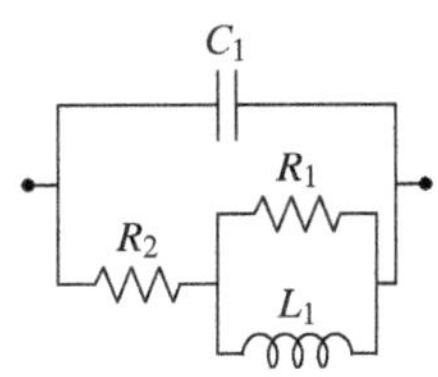

Network #49

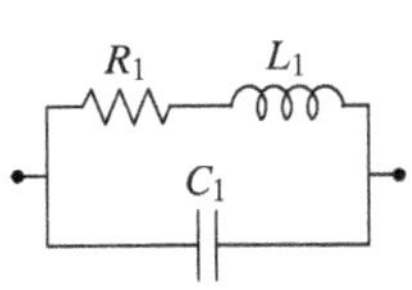

Network #50

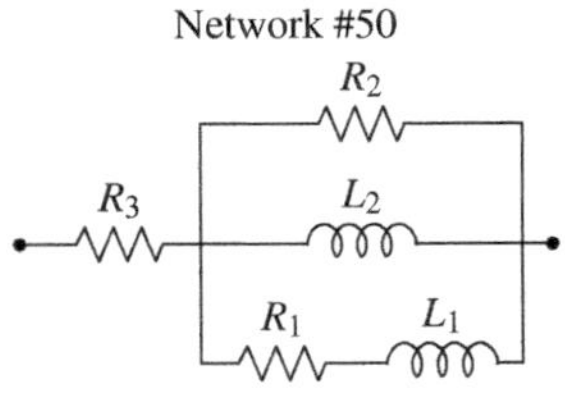

Network #51

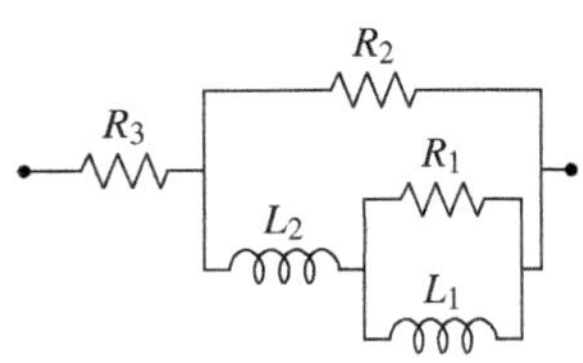

Network #52

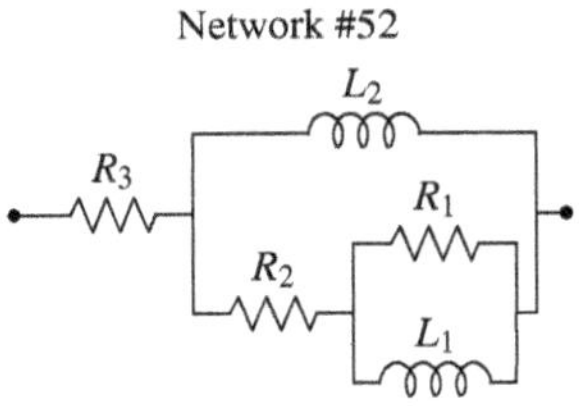

Network #53

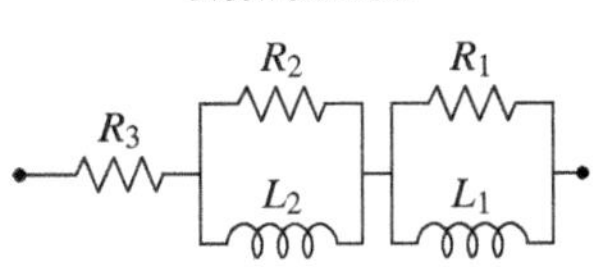

Network #54

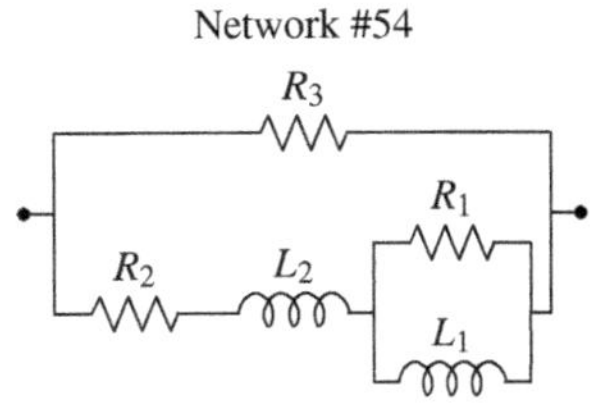

Network #55

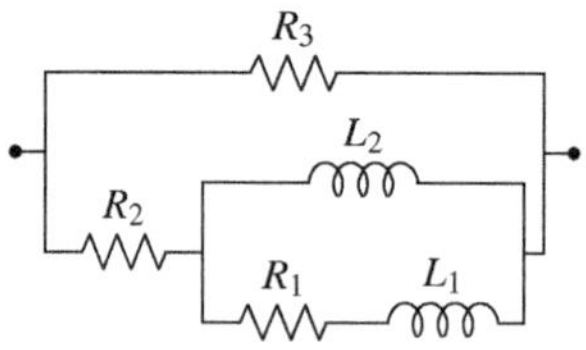

Network #56

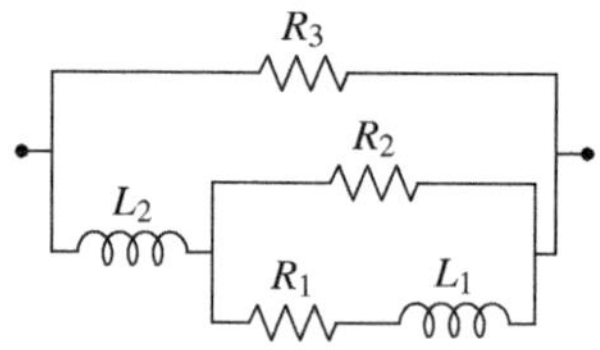

Network #57

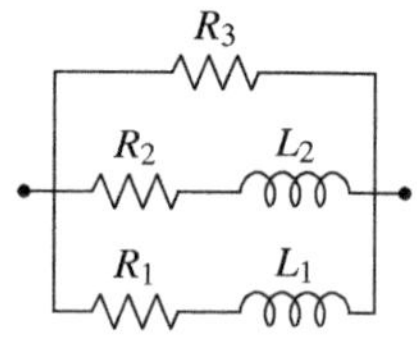

Network #58

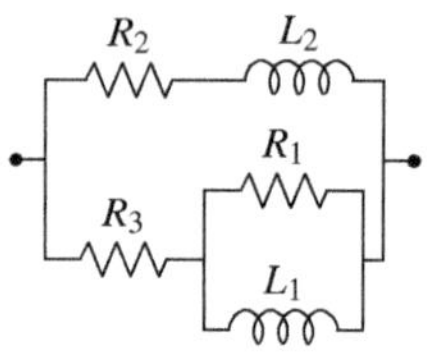

Network #59

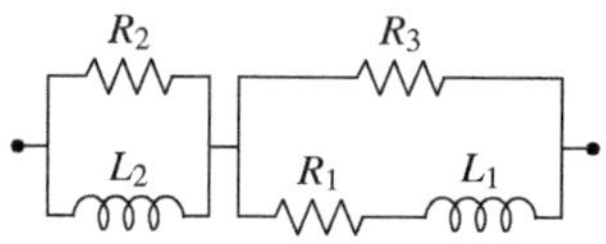

Network #60

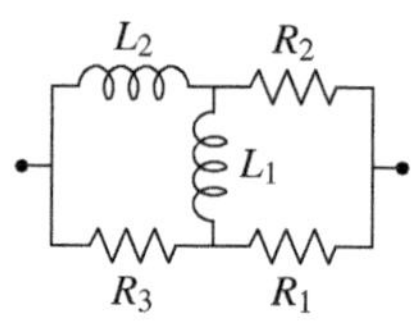

Network #61

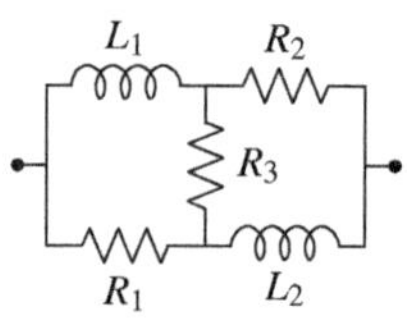

Network #62

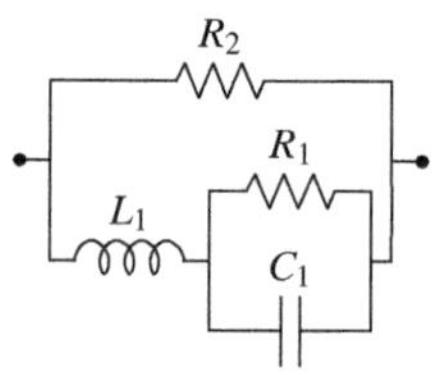

Network #63

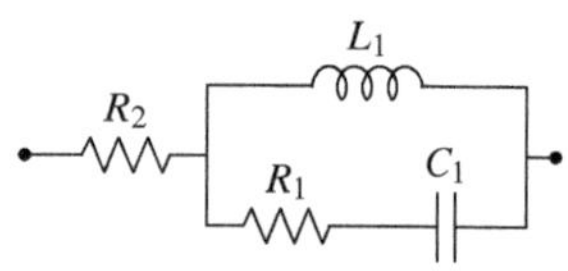

Network #64

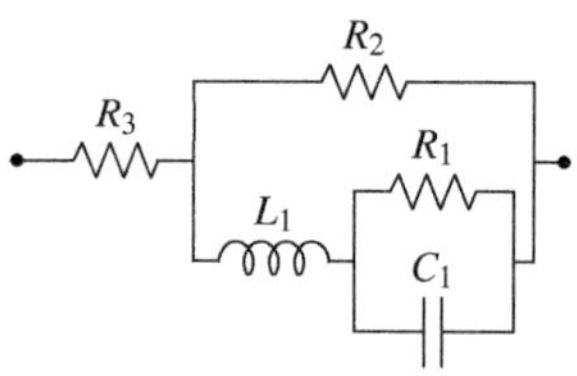

Network #65

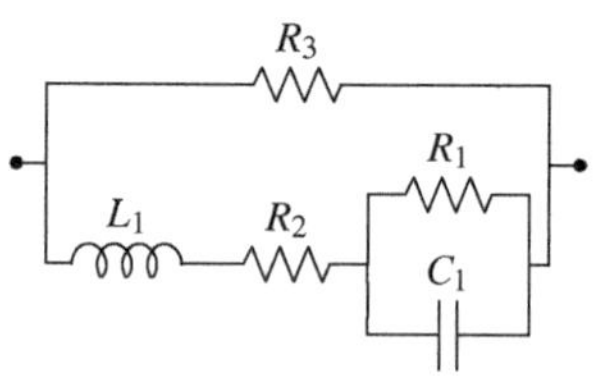

Network #66

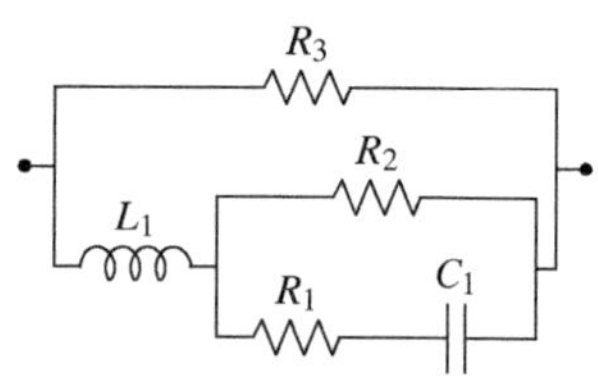

Network #67

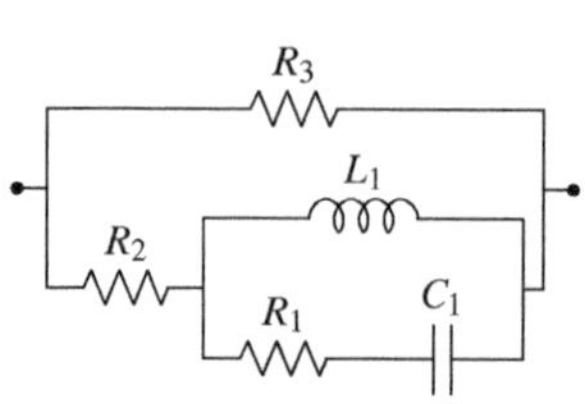

Network #68

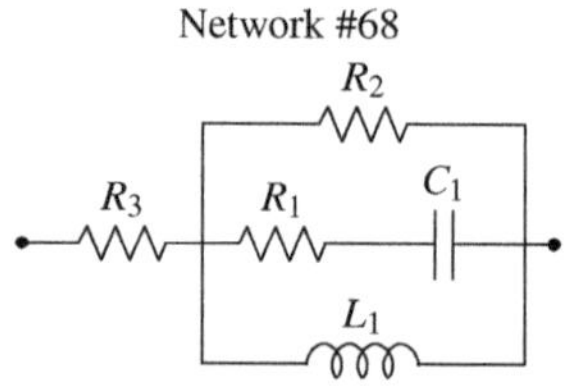

Network #69

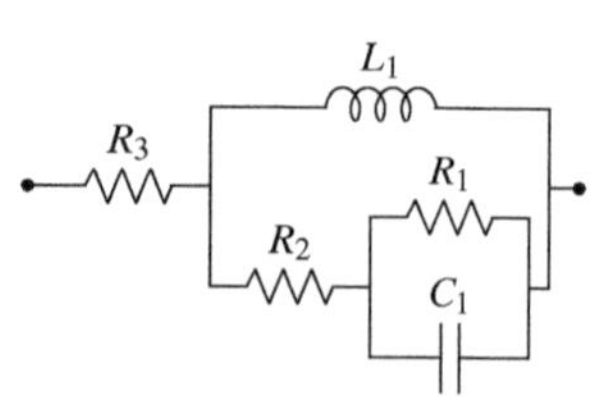

Network #70

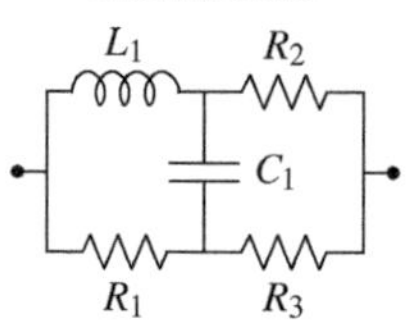

Network #71

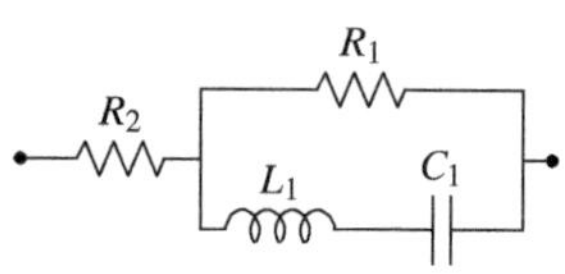

Network #72

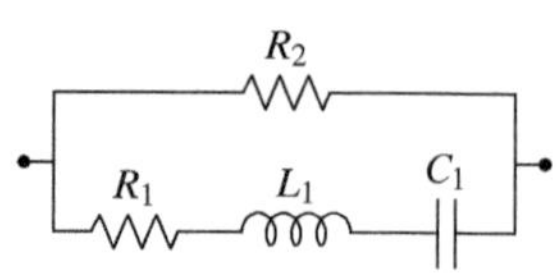

Network #73

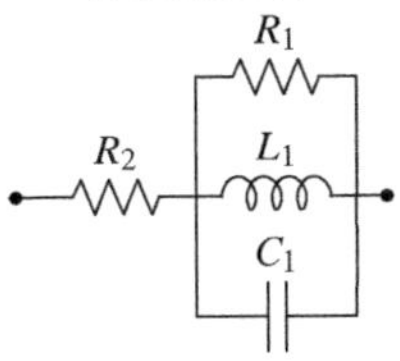

Network #74

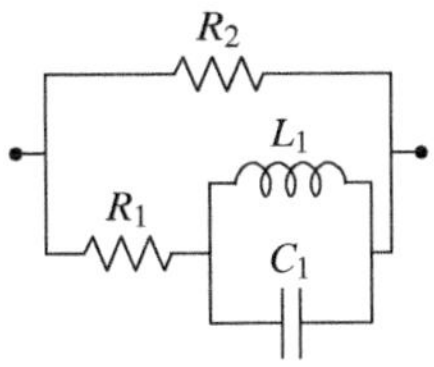

Network #75

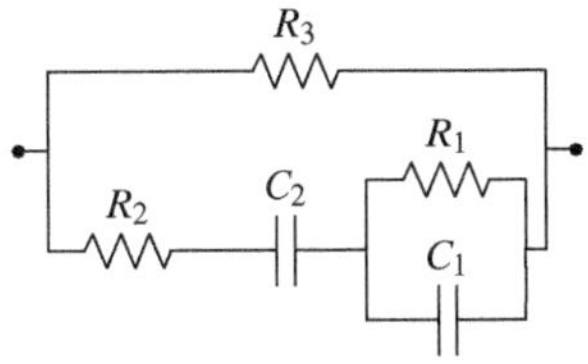

Network #76

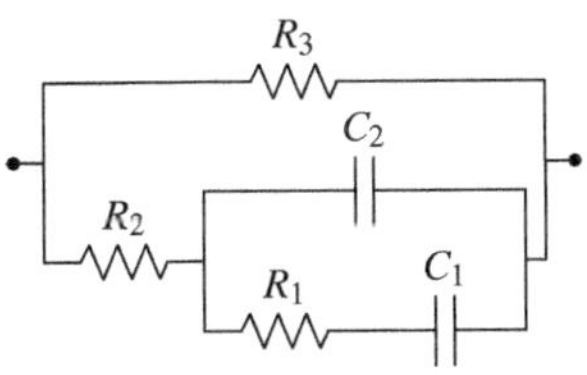

Network #77

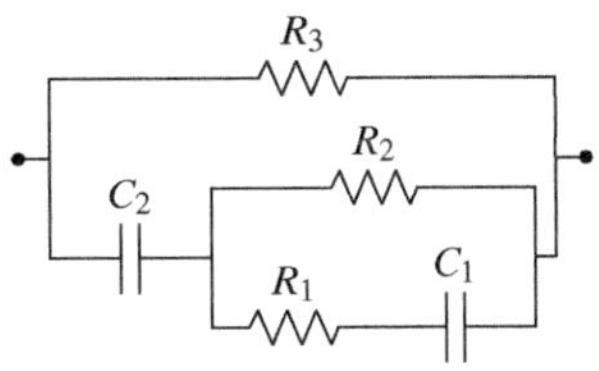

Network #78

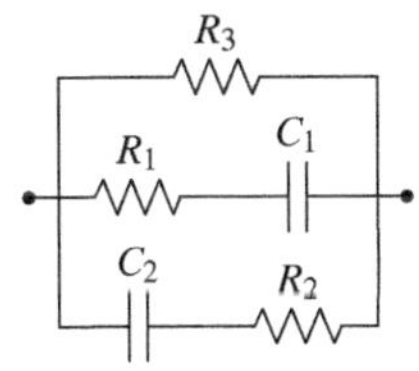

Network #79

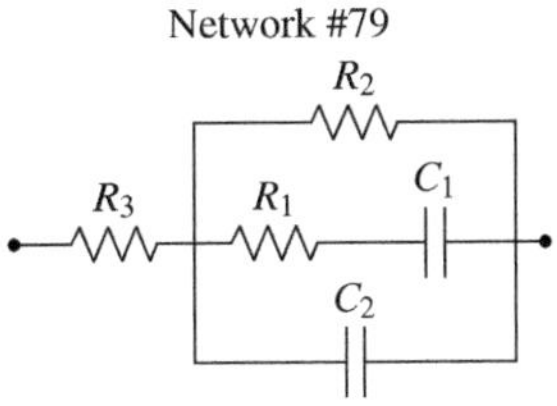

Network #80

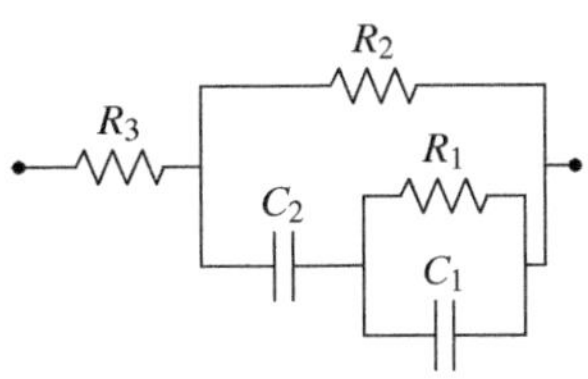

Network #81

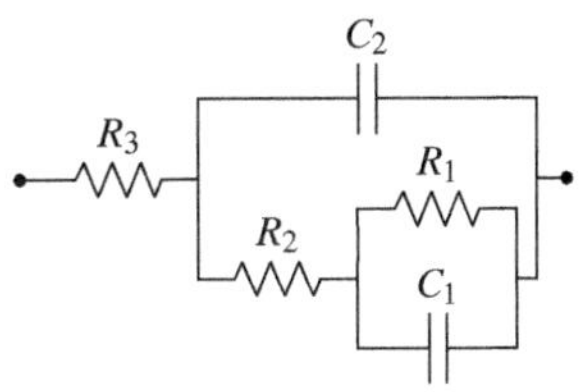

Network #82

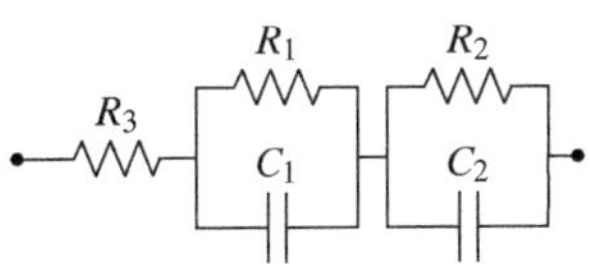

Network #83

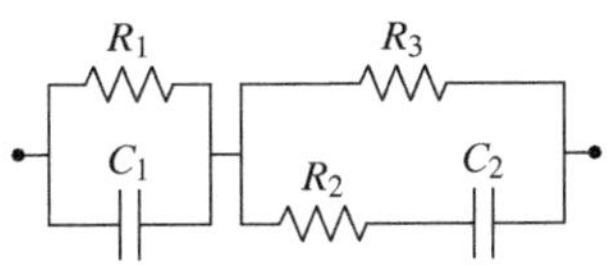

Network #84

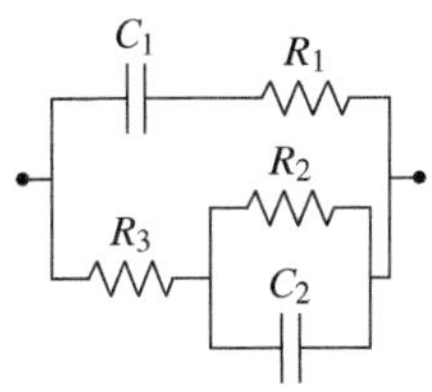

Network #85

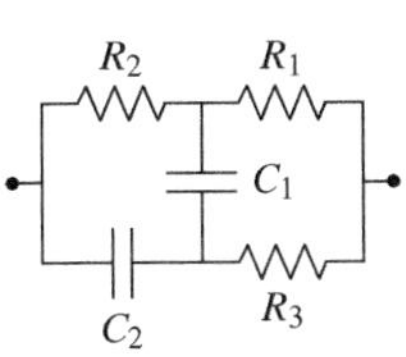

Network #86

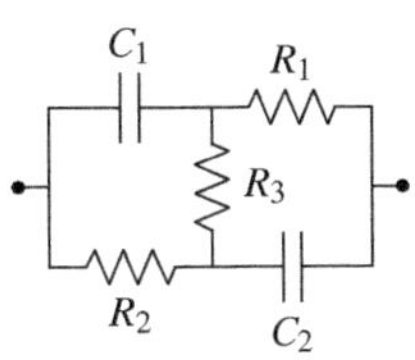

Network #87

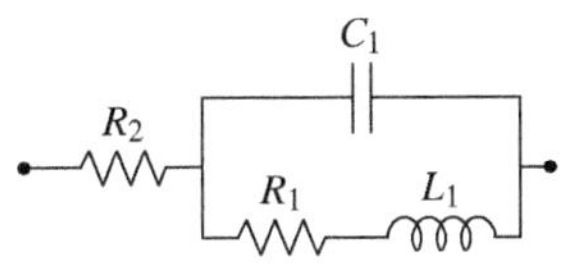

Network #88

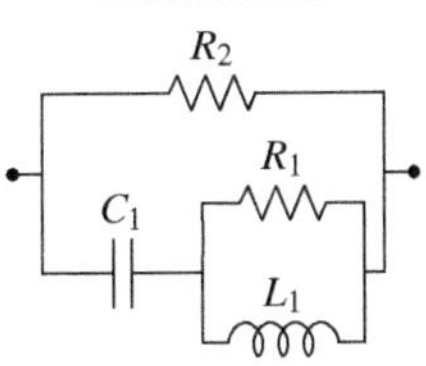

Network #89

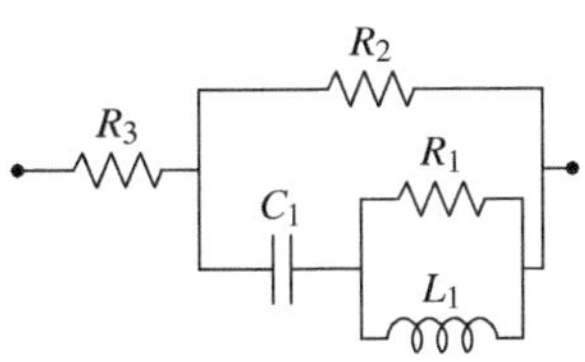

Network #90

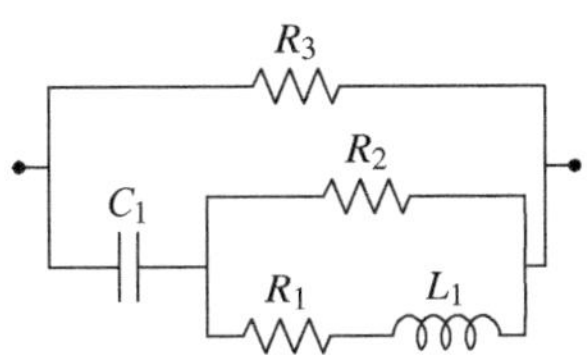

Network #91

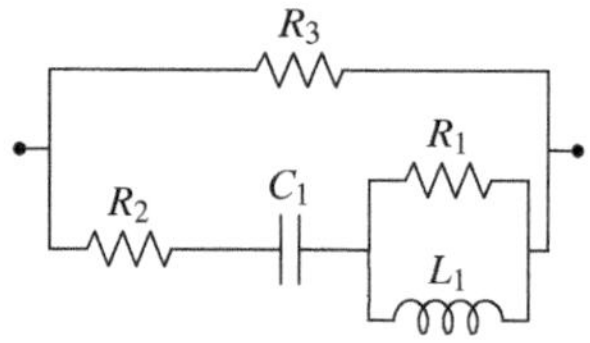

Network #92

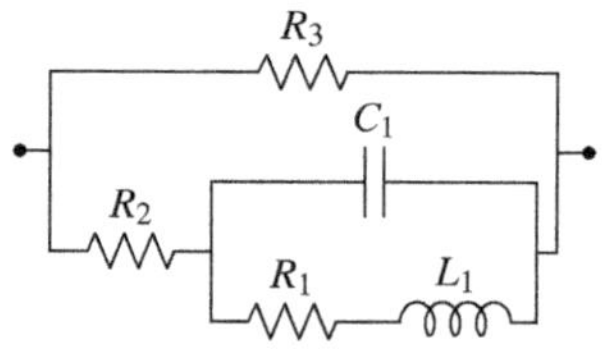

Network #93

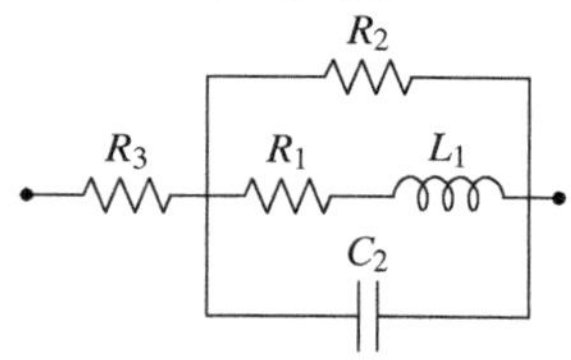

Network #94

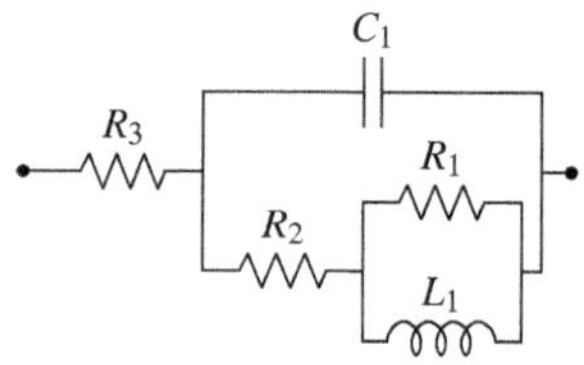

Network #95

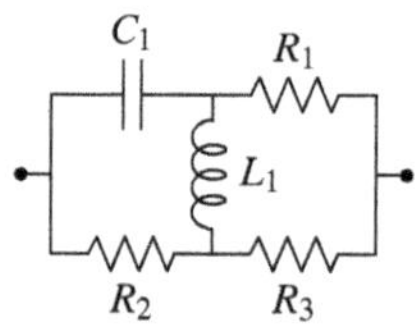

Network #96

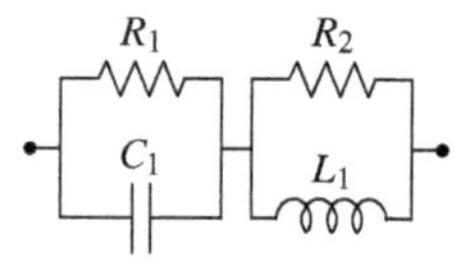

Network #97

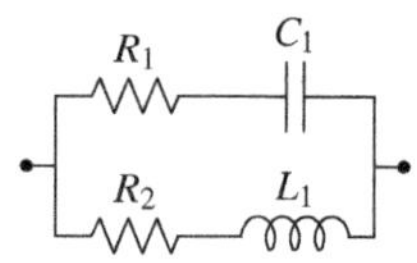

Network #98

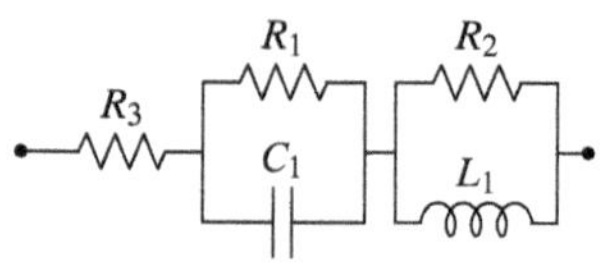

Network #99

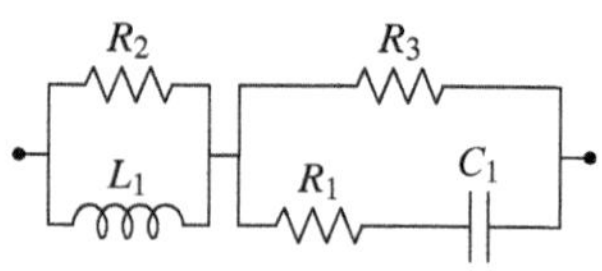

Network #100

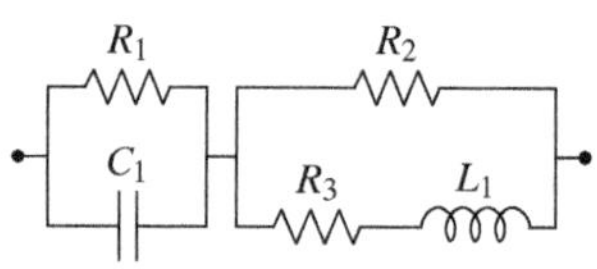

Network #101

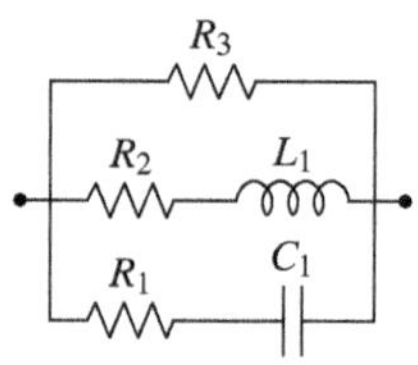

Network #102

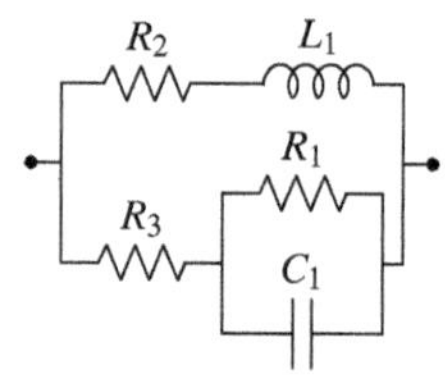

Network #103

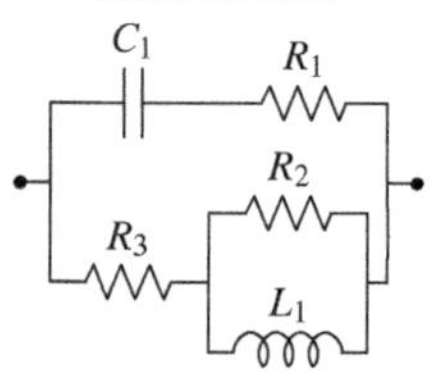

Network #104

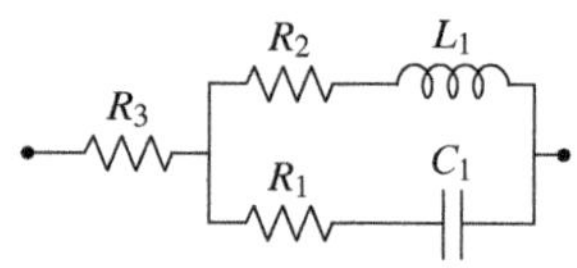

Network #105

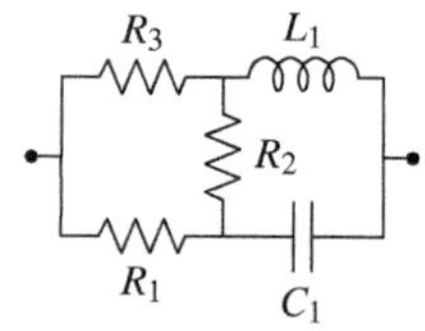

Network #106

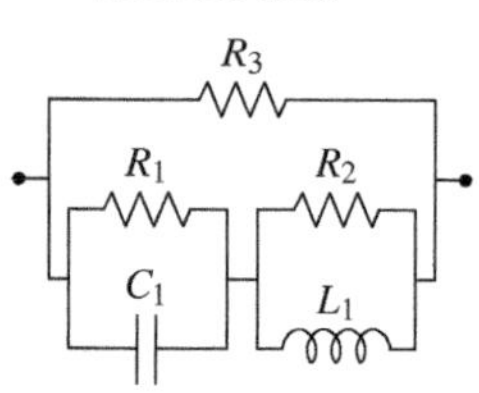

Network #107

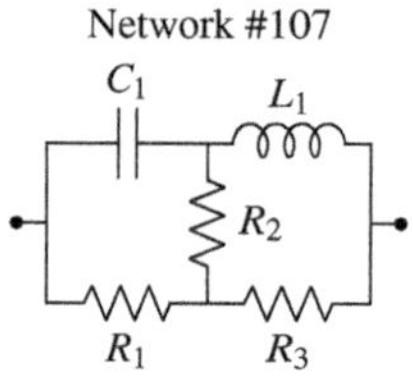

Network #108

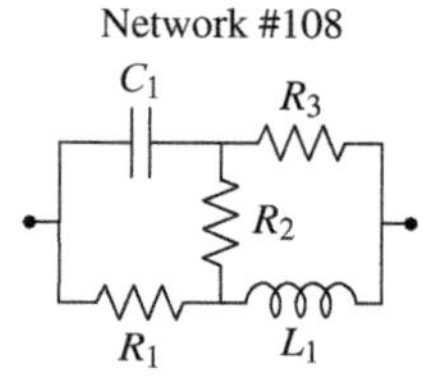

Appendix D

The Ladenheim networks (subfamily order)

Subf. I_A

Network #3

Subf. I_B

Network #1

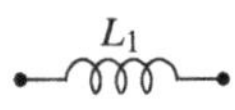

Network #2

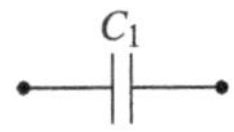

Subf. II_A

Network #6

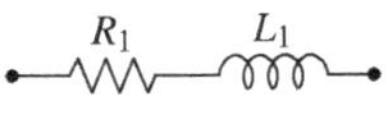

Network #8

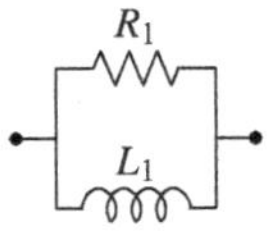

Network #5

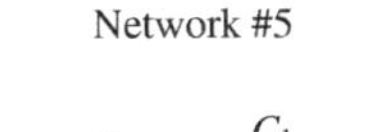

Network #9

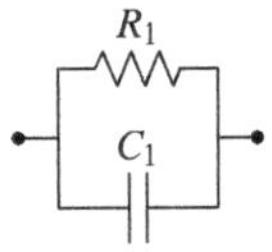

Subf. II_B

Network #4

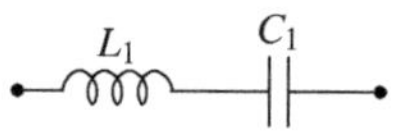

Network #7

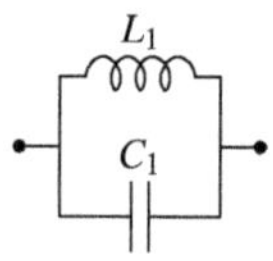

Subf. III_A

Network #15

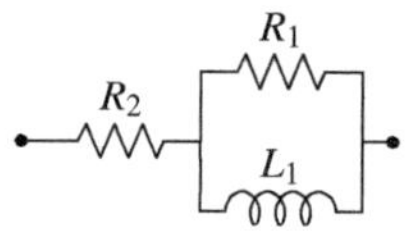

Network #16

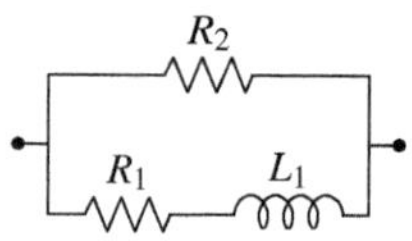

Network #17

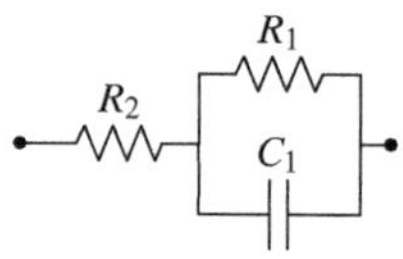

Network #18

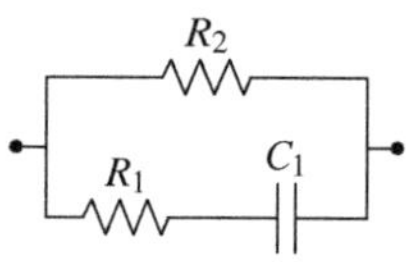

Subf. III_B

Network #11

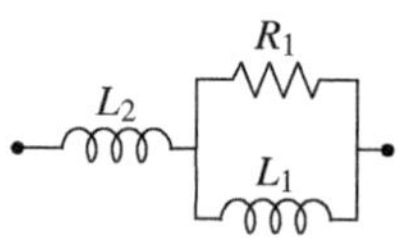

Network #12

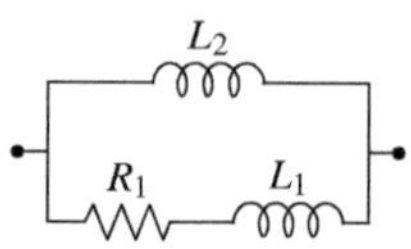

Network #14

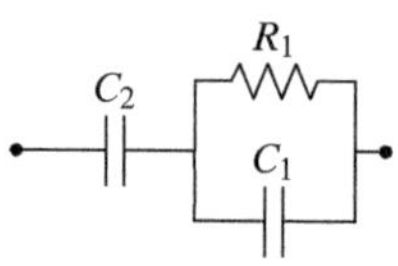

Network #13

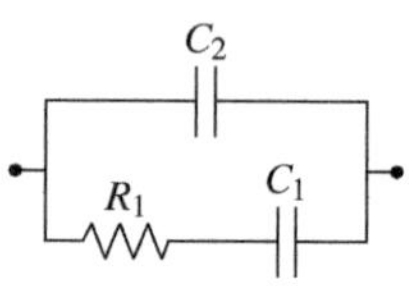

Subf. III_C

Network #41

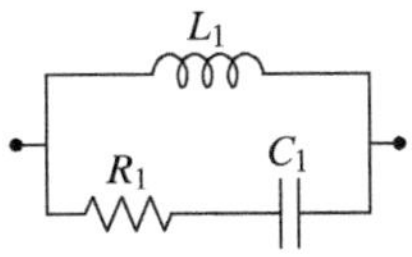

Network #34

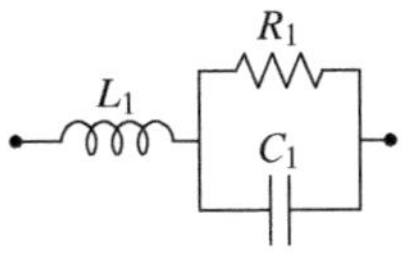

Network #49

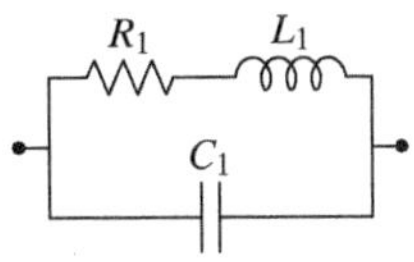

Network #26

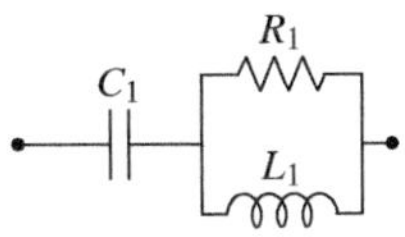

Subf. III_D

Network #27

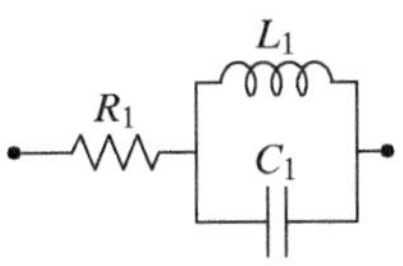

Network #42

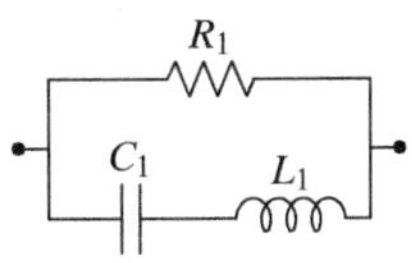

Subf. III_E

Network #10

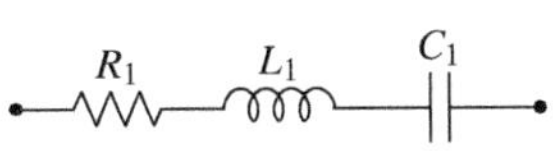

Network #19

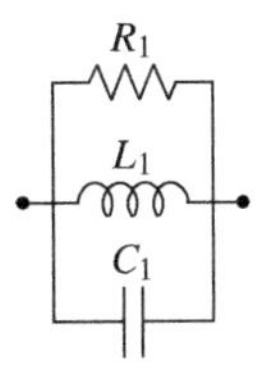

Subf. IV_A

Network #37

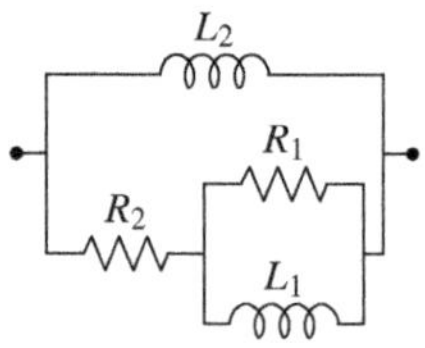

Network #30

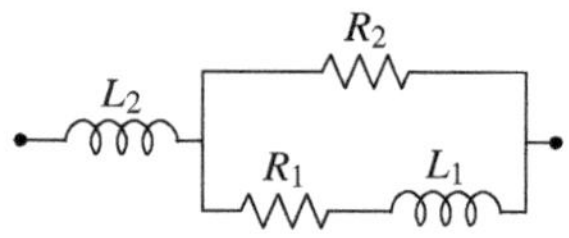

Network #35

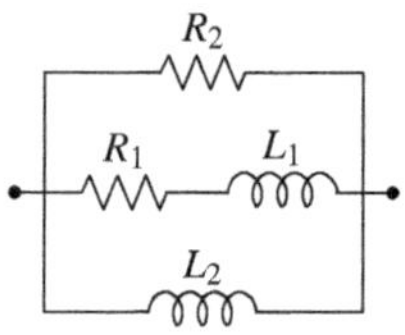

Network #28

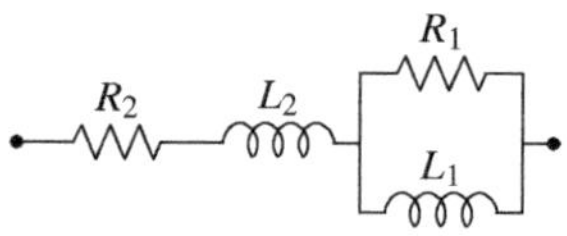

Network #36

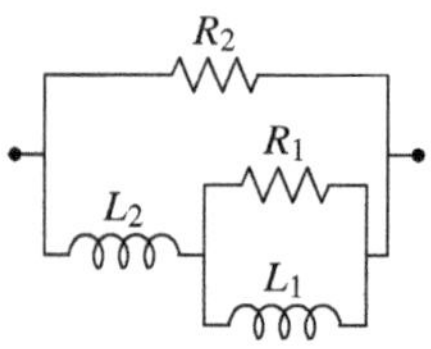

Network #29

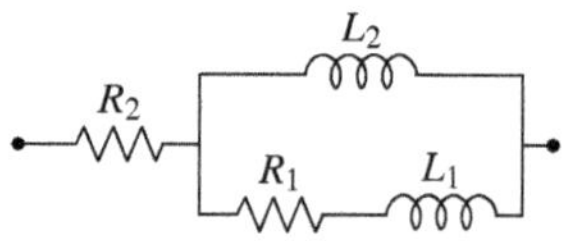

Network #44

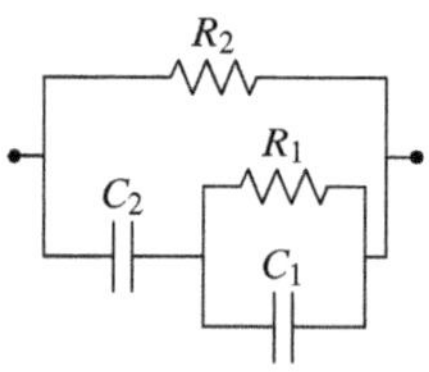

Network #21

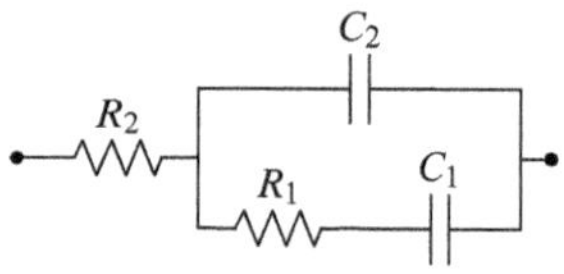

Network #43

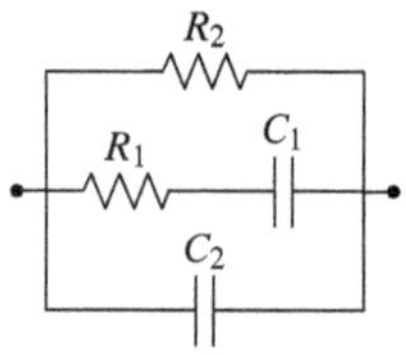

Network #20

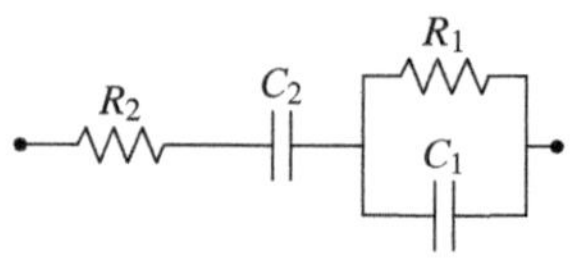

Network #45

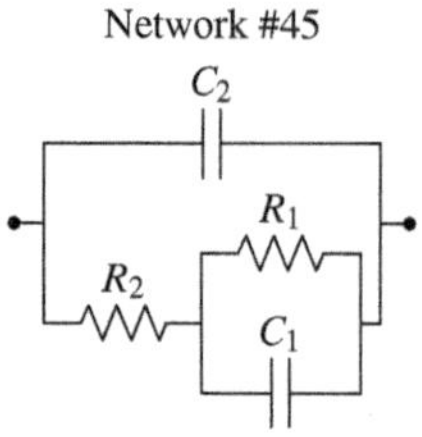

Network #22

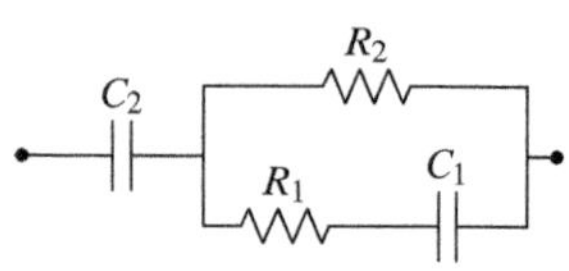

Subf. IV_B

Network #38

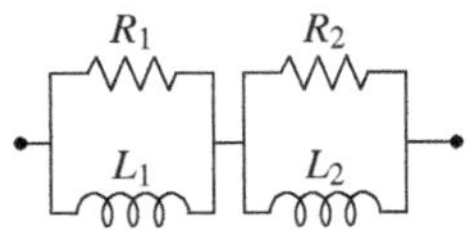

Network #31

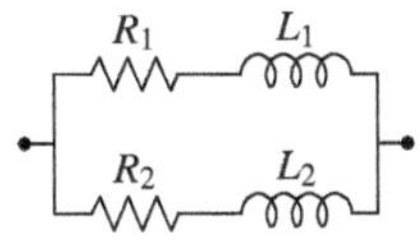

Network #46

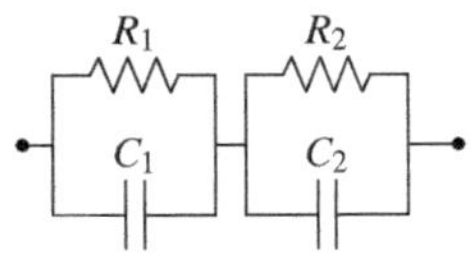

Network #23

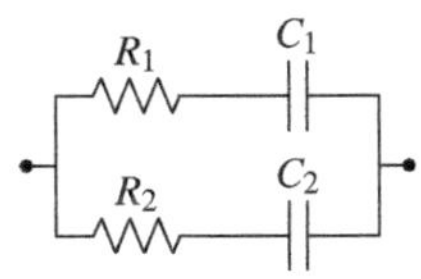

Subf. IV_C

Network #40

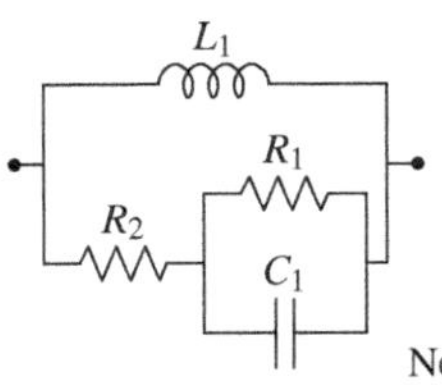

Network #33

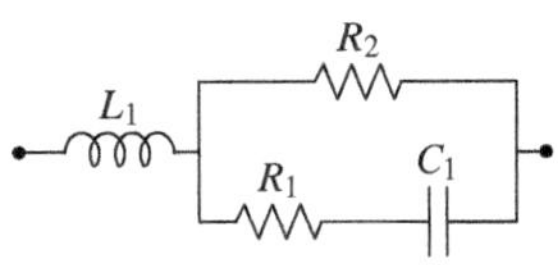

Network #39

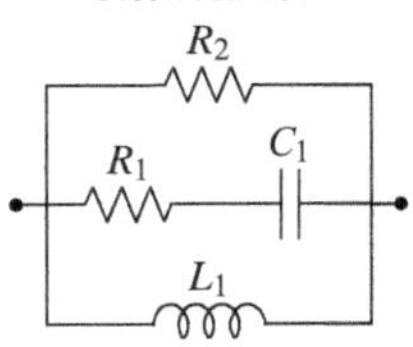

Network #32

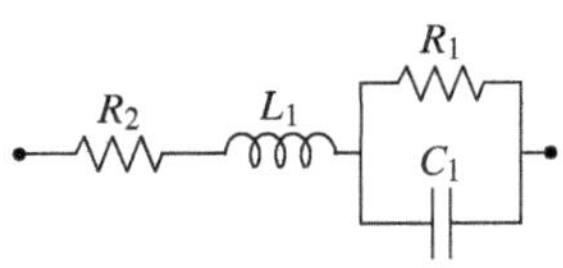

Network #47

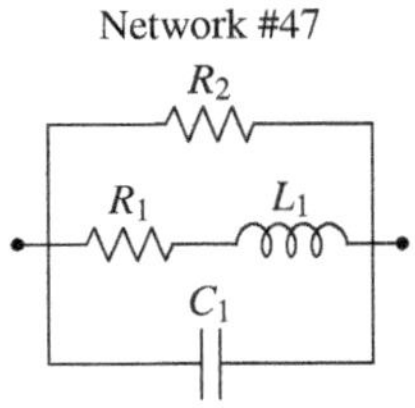

Network #25

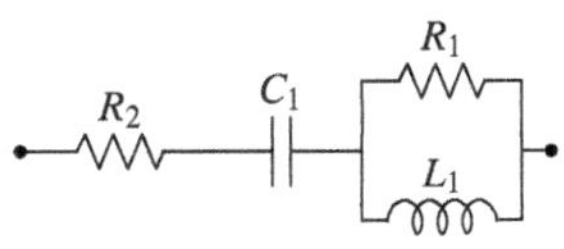

Network #48

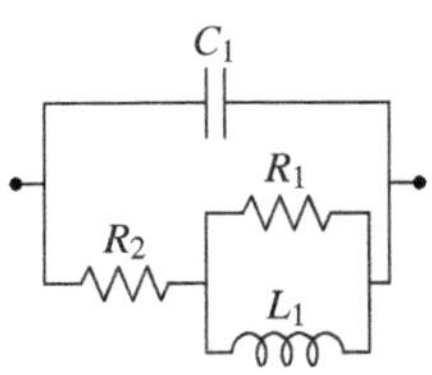

Network #24

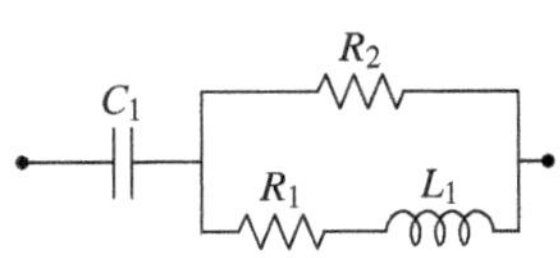

Subf. IV_D

Network #72

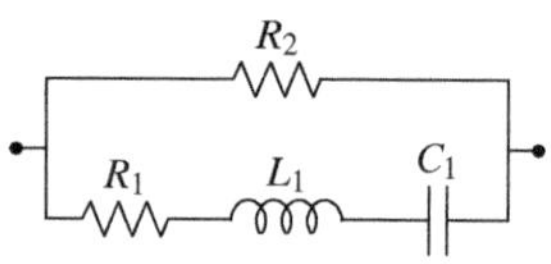

Network #73

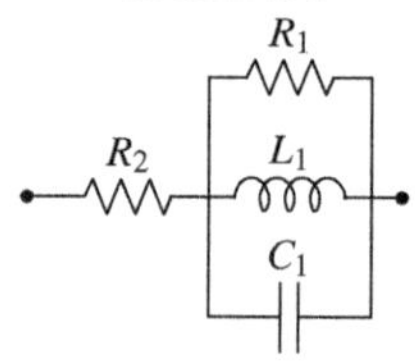

Network #71

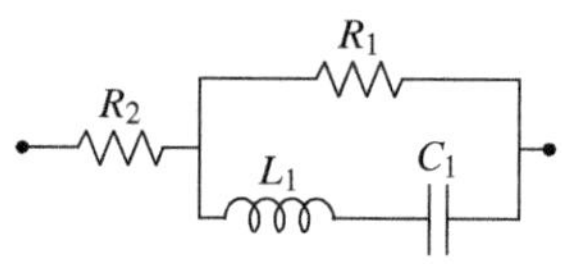

Network #74

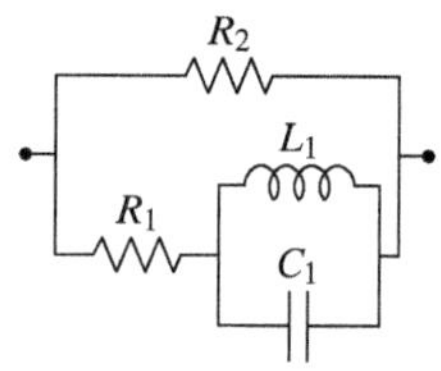

Subf. IV_E

Network #63

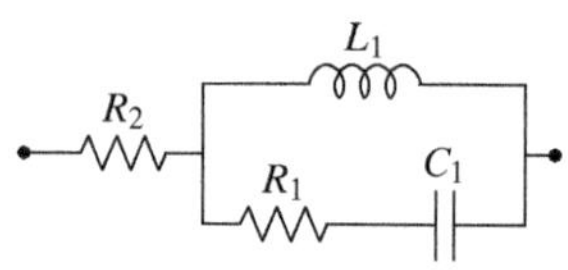

Network #62

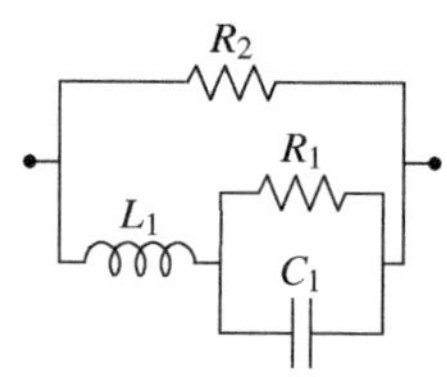

Network #87

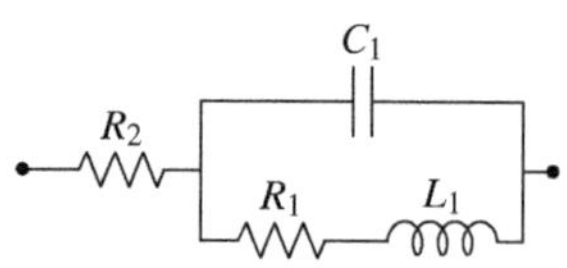

Network #88

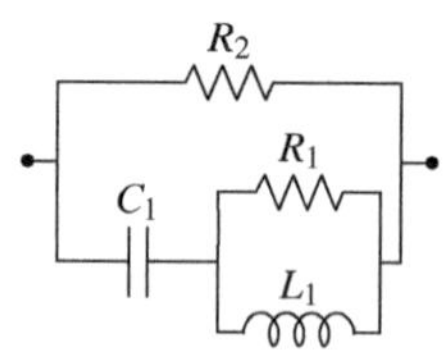

Subf. IV_F

Network #97

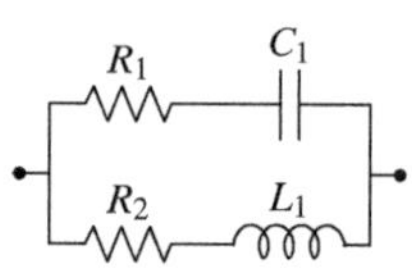

Network #96

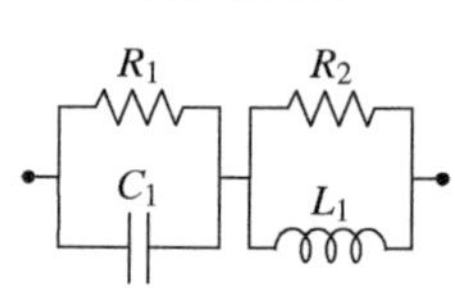

Subf. V_A

Network #52

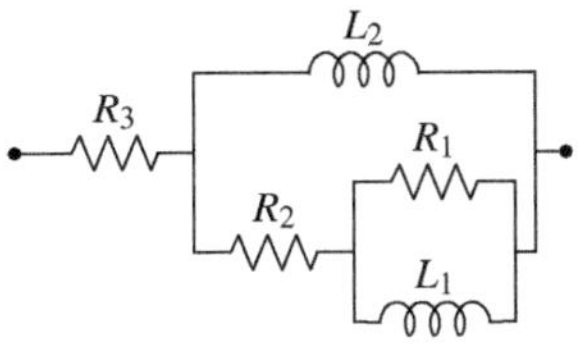

Network #56

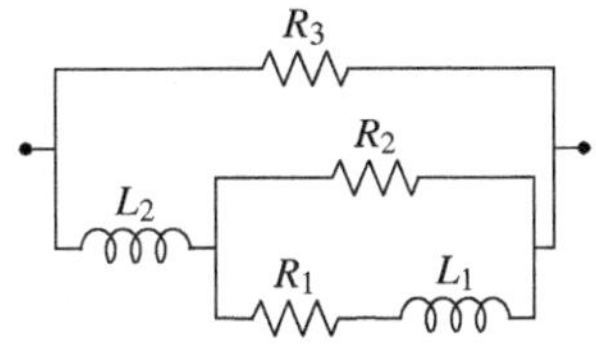

Network #50

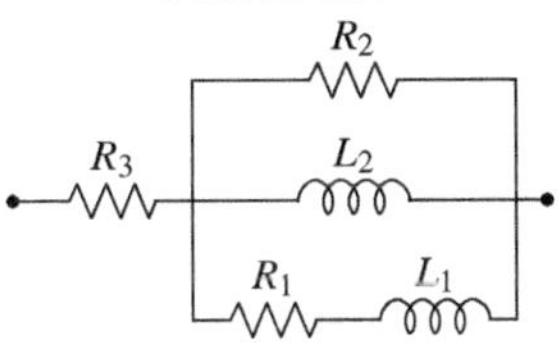

Network #54

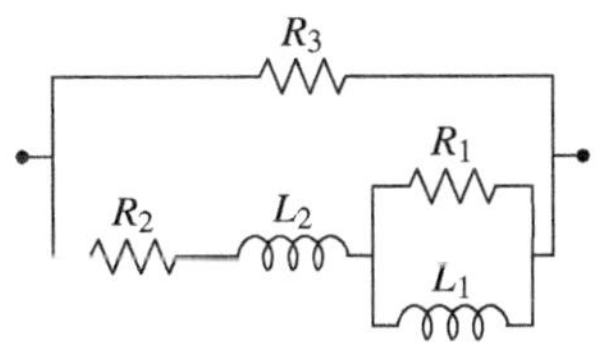

Network #51

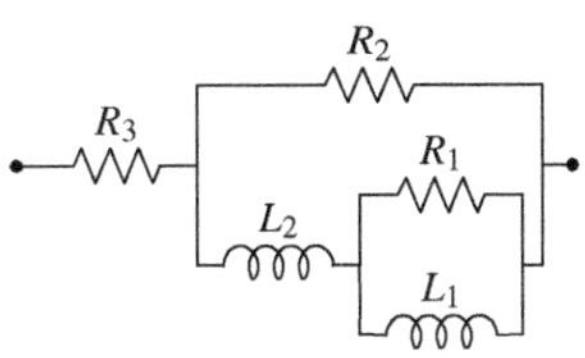

Network #55

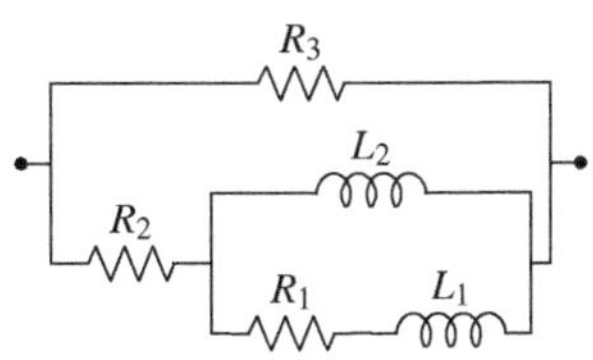

Network #80

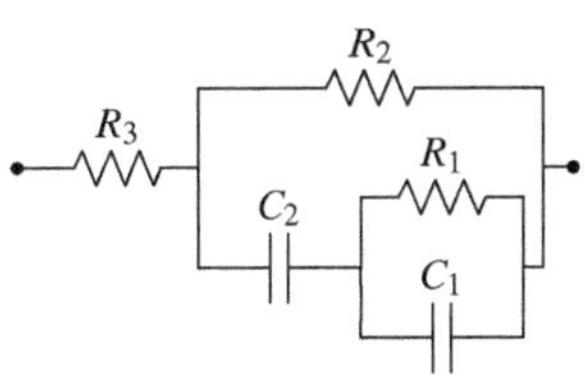

Network #76

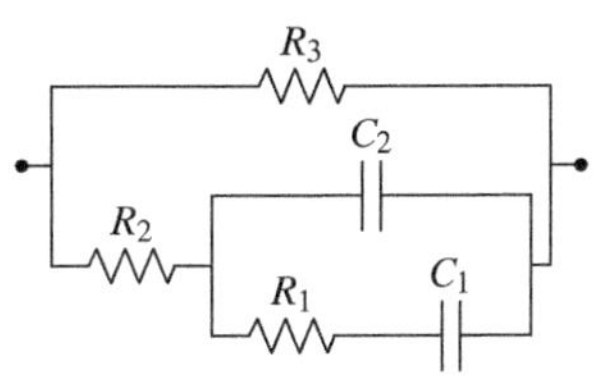

Network #79

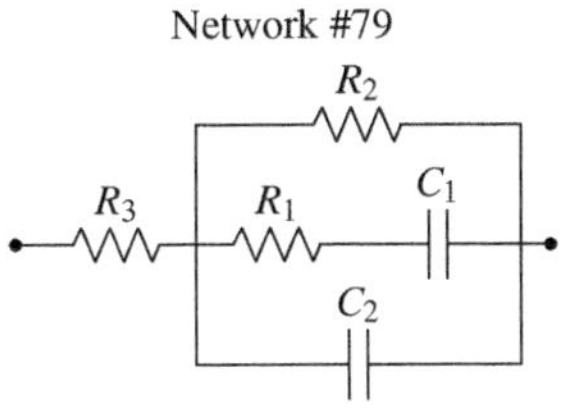

Network #75

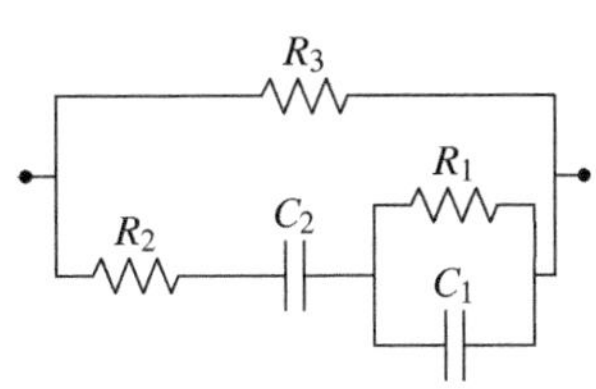

Network #81

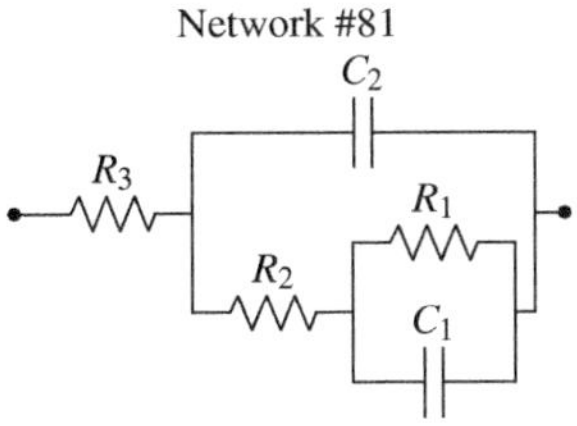

Network #77

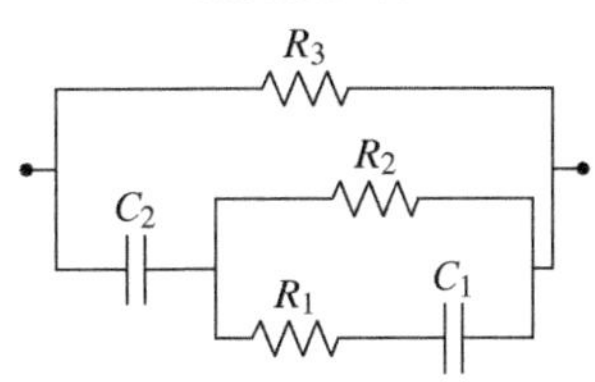

Subf. V_B

Network #59

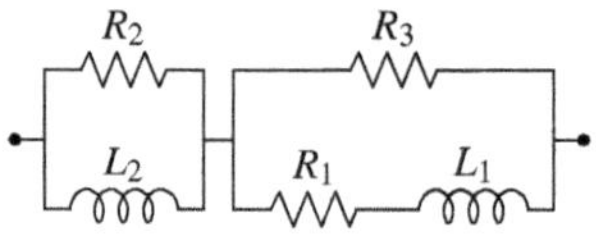

Network #58

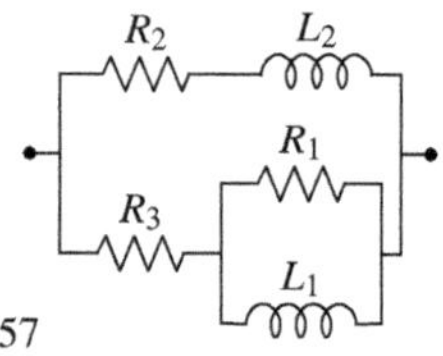

Network #53

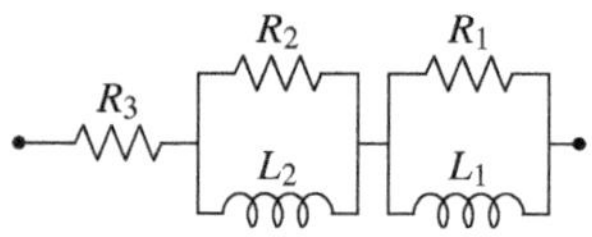

Network #57

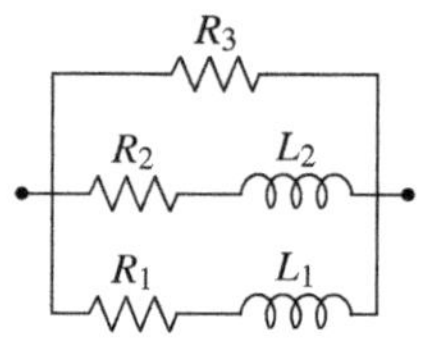

Network #82

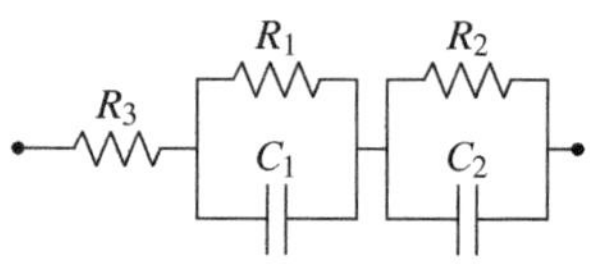

Network #78

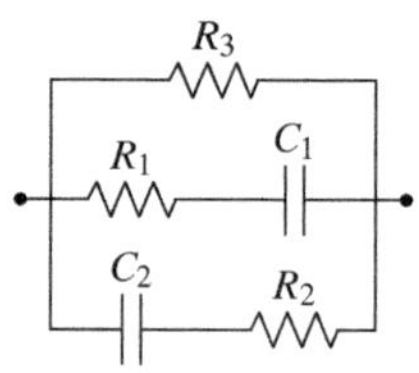

Network #83

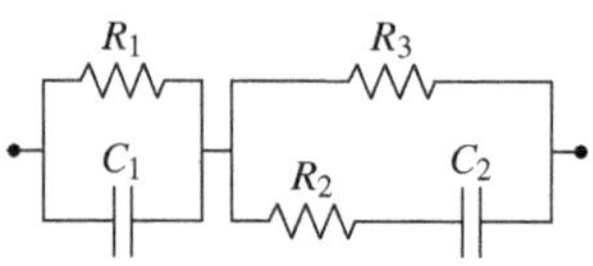

Network #84

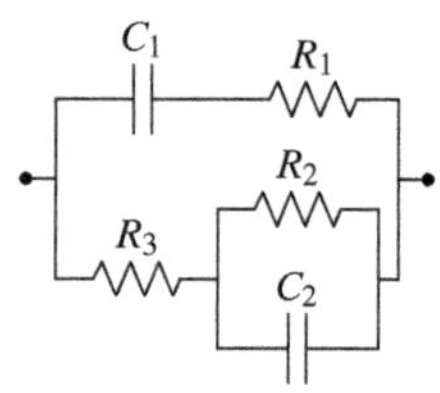

Subf. V_C

Network #60

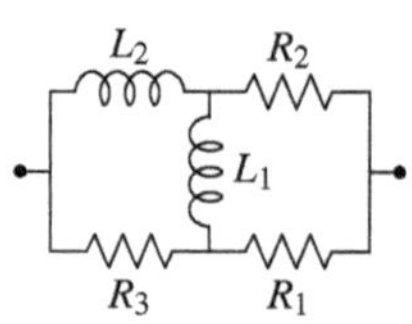

Network #85

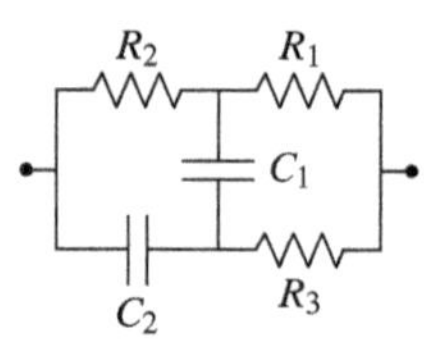

Subf. V_D

Network #61

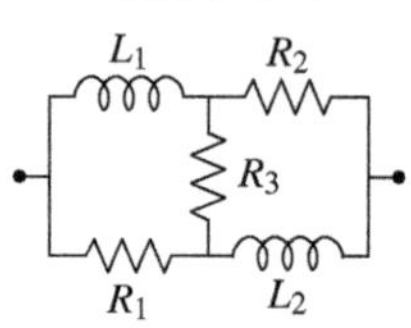

Network #86

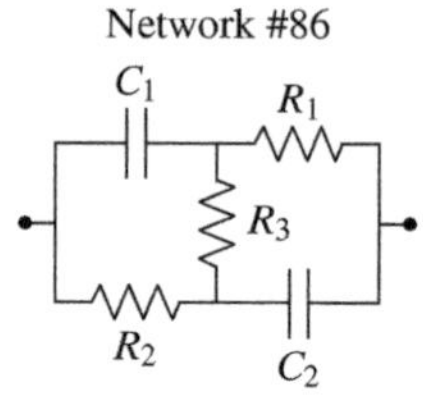

Subf. V_E

Network #69

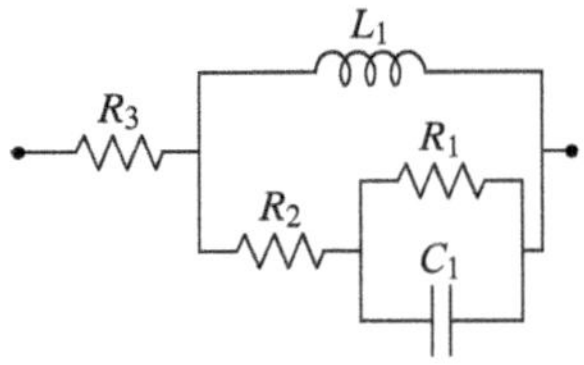

Network #66

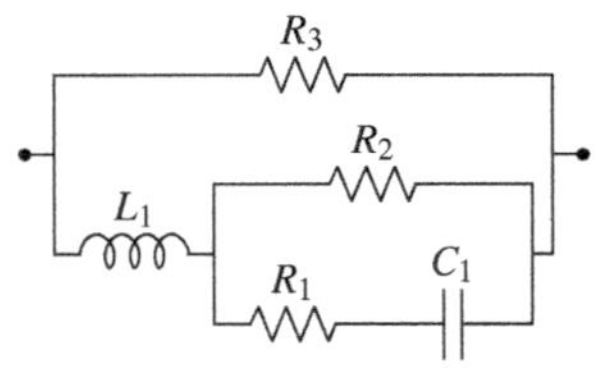

Network #68

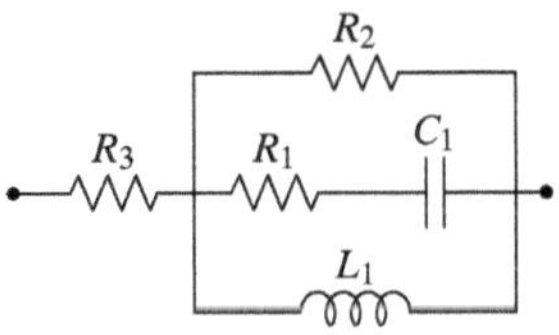

Network #65

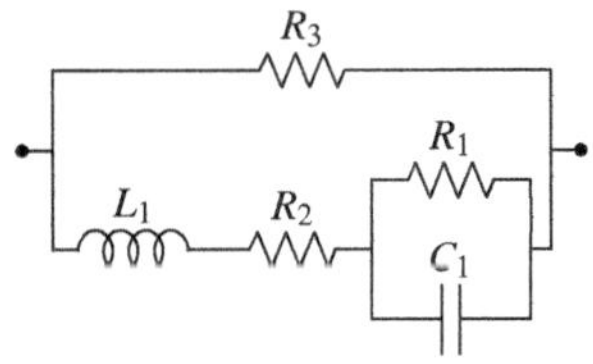

Network #67

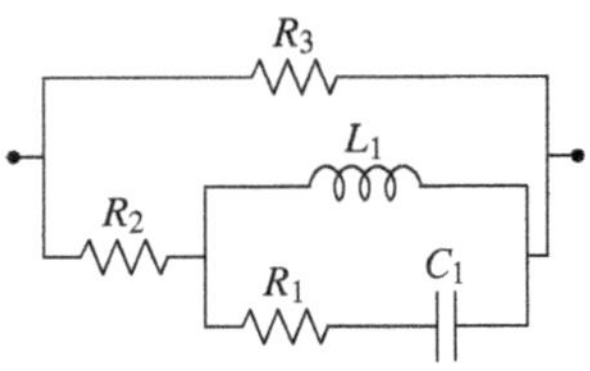

Network #64

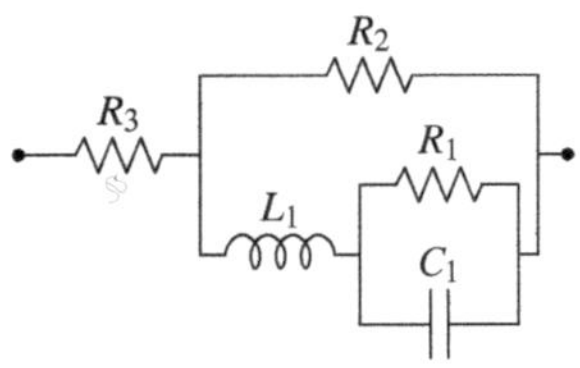

Network #92

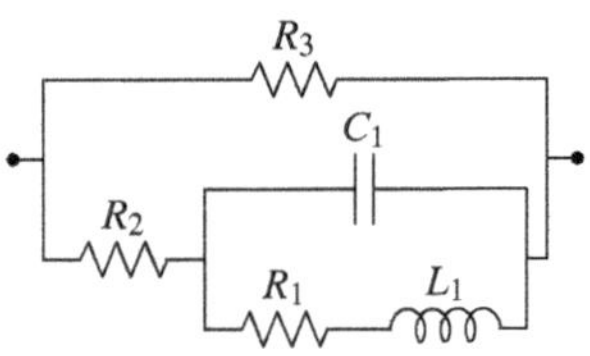

Network #89

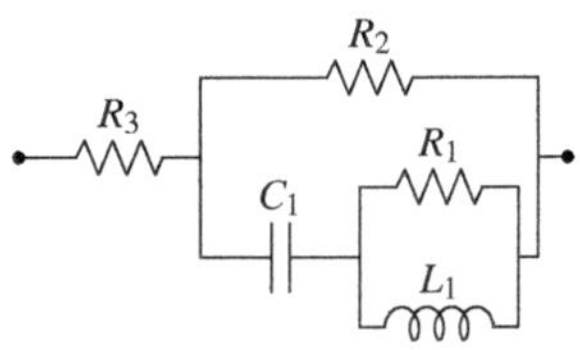

Network #93

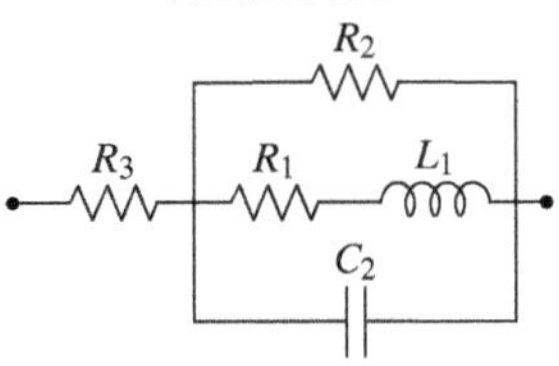

Network #91

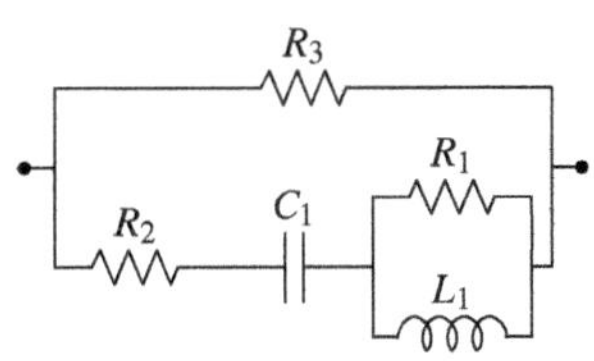

Network #94

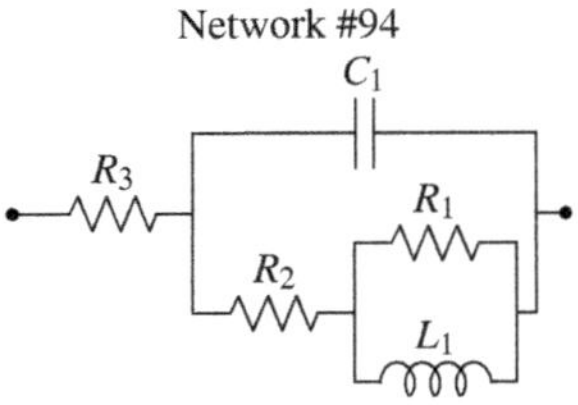

Network #90

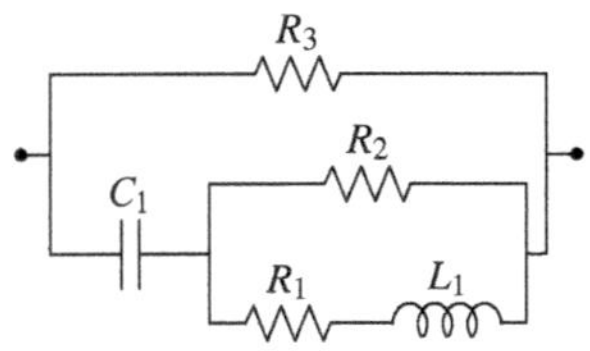

Subf. V_F

Network #102

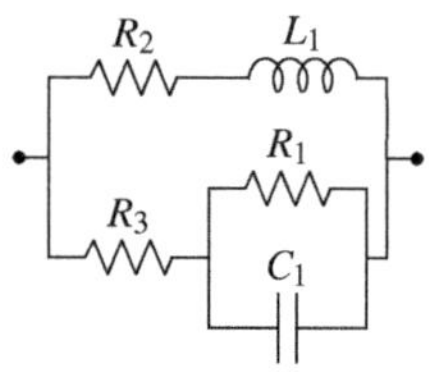

Network #99

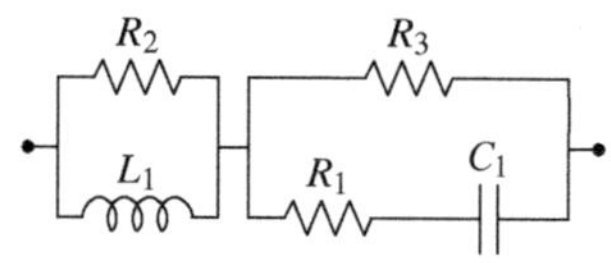

Network #101

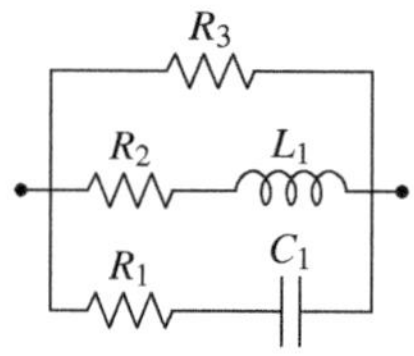

Network #98

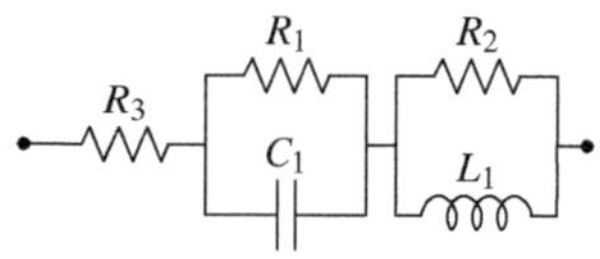

Network #103

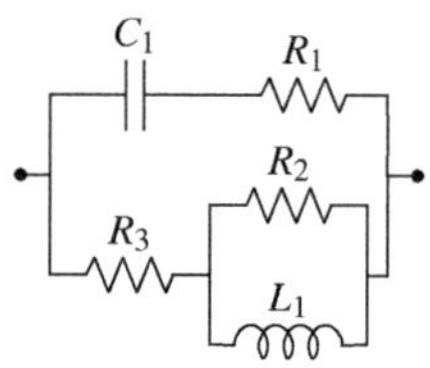

Network #100

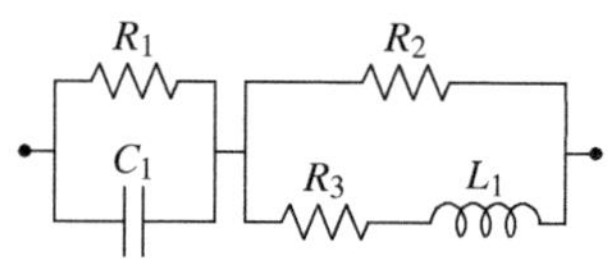

Subf. V_G

Network #104

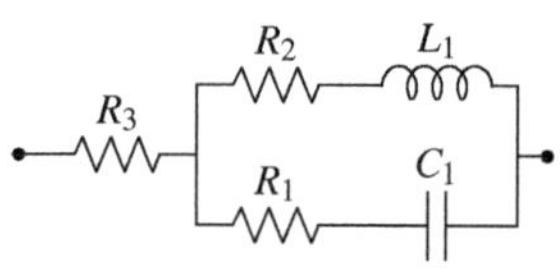

Network #106

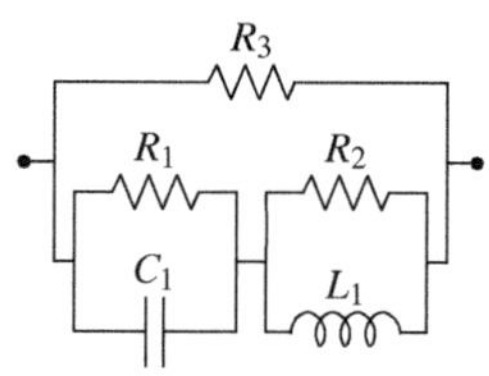

Network #105

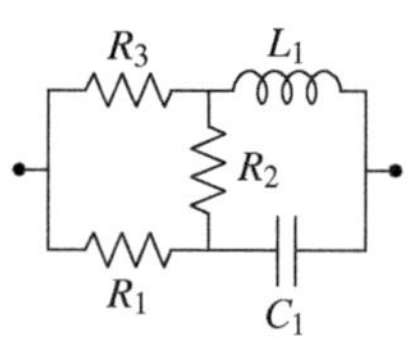

Network #107

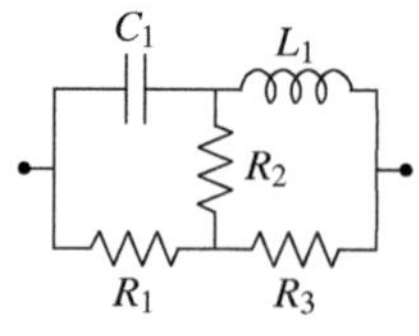

Subf. V_H

Network #70

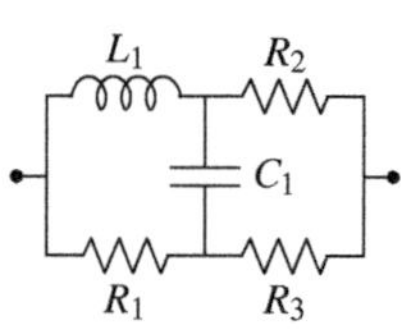

Network #95

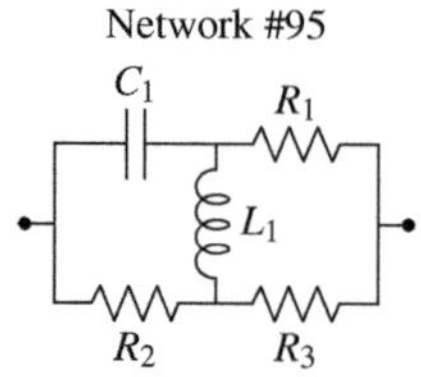

Subf. V_I

Network #108

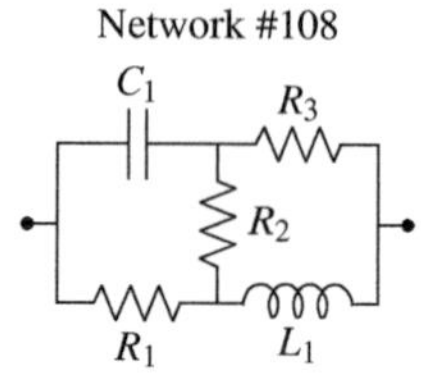

Bibliography

[1] B. D. O. Anderson and S. Vongpanitlerd, *Network Analysis and Synthesis*, Prentice–Hall, Englewood Cliffs, NJ, 1st ed., 1973. (Cited on pp. 5, 10, 11, 16, 17, 18)

[2] L. V. Auth, *RLC biquadratic driving-point synthesis using the resistive three-port*, IEEE Trans. Circuit Theory, 11 (1964), pp. 82–88. (Cited on pp. 18, 24)

[3] S. Basu, R. Pollack, and M.-F. Roy, *Algorithms in Real Algebraic Geometry*, Springer, Berlin, Heidelberg, 2nd ed., 2006. (Cited on p. 70)

[4] A. Beardon, *Algebra and Geometry*, Cambridge University Press, Cambridge, UK, 1st ed., 2005. (Cited on p. 45)

[5] E. Bierstone and P. D. Milman, *Semianalytic and subanalytic sets*, Publ. Math. Inst. Hautes Études Sci., 67 (1988), pp. 5–42. (Cited on p. 44)

[6] G. Birkhoff, *On the structure of abstract algebras*, Math. Proc. Cambridge Philos. Soc., 31 (1935), pp. 433–454. (Cited on p. 72)

[7] R. Bott and R. J. Duffin, *Impedance synthesis without use of transformers*, J. Appl. Phys., 20 (1949), p. 816. (Cited on pp. 1, 13)

[8] O. Brune, *Synthesis of a finite two-terminal network whose driving-point impedance is a prescribed function of frequency*, J. Math. Phys., 10 (1931), pp. 191–236. (Cited on pp. 1, 9, 10, 11)

[9] S. Bruni, R. Goodall, T. X. Mei, and H. Tsunashima, *Control and monitoring for railway vehicle dynamics*, Vehicle Syst. Dyn., 45 (2007), pp. 743–779. (Cited on p. 36)

[10] W. Cauer, *Die Verwirklichung von Wechselstromwiderständen vorgeschriebener Frequenzabhängigkeit (The realisation of impedances of prescribed frequency dependence)*, Arch. Elektrotechnik, 17 (1926), pp. 355–388. (Cited on pp. 1, 8, 67)

[11] ———, *Ein Reaktanztheorem*, Sitzungsber. Preuss. Akad. Wiss. Physik. Math. Kl., 30 (1931), pp. 673–681. (Cited on p. 16)

[12] ———, *Synthesis of Linear Communication Networks*, McGraw–Hill, New York, 2nd ed., 1958. (Cited on p. 5)

[13] I. Cederbaum, *Conditions for the impedance and admittance matrices of n-ports without ideal transformers*, in Proceedings of the IEE - Part C: Monographs, vol. 105, 1958, pp. 245–251. (Cited on p. 18)

[14] H. J. Chen, W. J. Su, and F. C. Wang, *Modeling and analyses of a connected multicar train system employing the inerter*, Adv. Mech. Eng., 9 (2017), pp. 1–13. (Cited on p. 36)

[15] M. Z. Q. Chen, C. Papageorgiou, F. Scheibe, F. C. Wang, and M. C. Smith, *The missing mechanical circuit element*, IEEE Circuits Syst. Mag., 9 (2009), pp. 10–26. (Cited on pp. 29, 33)

[16] M. Z. Q. Chen and M. C. Smith, *Electrical and mechanical passive network synthesis*, in Recent Advances in Learning and Control, Lect. Notes Control Inf. Sci. 371, Springer, London, 2008, pp. 35–50. (Cited on p. 23)

[17] ——, *A note on tests for positive-real functions*, IEEE Trans. Automat. Control, 54 (2009), pp. 390–393. (Cited on p. 21)

[18] M. Z. Q. Chen, K. Wang, C. Li, and G. Chen, *Realization of biquadratic impedances as five-element bridge networks*, IEEE Trans. Circuits Syst., 64 (2017), pp. 1599–1611. (Cited on pp. 20, 56, 73, 98)

[19] S. Darlington, *Synthesis of reactance 4-poles which produce prescribed insertion loss characteristics*, J. Math. Phys., 18 (1939), pp. 257–353. (Cited on pp. 1, 15)

[20] M. Darrieus, *Les modèles mécaniques en électrotechnique, leur application aux problèmes de stabilité*, Bull. Soc. Fra. Elect., 96 (1929), pp. 794–809. (Cited on p. 29)

[21] S. Evangelou, D. J. N. Limebeer, R. S. Sharp, and M. C. Smith, *Control of motorcycle steering instabilities*, IEEE Control Syst. Mag., 26 (2006), pp. 78–88. (Cited on pp. 32, 35)

[22] ——, *Mechanical steering compensators for high-performance motorcycles*, J. Appl. Mech., 74 (2007), pp. 332–346. (Cited on pp. 32, 35)

[23] A. Fialkow and I. Gerst, *Impedance synthesis without mutual coupling*, Quart. Appl. Math., 12 (1955), pp. 420–422. (Cited on pp. 2, 14)

[24] F. A. Firestone, *A new analogy between mechanical and electrical systems*, J. Acoust. Soc. Am., 4 (1933), pp. 249–267. (Cited on pp. 29, 31)

[25] R. M. Foster, *A reactance theorem*, Bell Syst. Tech. J., 3 (1924), pp. 259–267. (Cited on pp. 1, 7, 67)

[26] ——, *Geometrical circuits of electrical networks*, Trans. Amer. Inst. Electr. Engrg., 51 (1932), pp. 309–317. (Cited on pp. 42, 79)

[27] ——, *Passive network synthesis*, in Proceedings of the Symposium on Modern Network Synthesis, Polytechnic Institute of Brooklyn, 1955, pp. 3–9. (Cited on p. 2)

[28] ——, *Academic and theoretical aspects of circuit theory*, Proc. IRE, 50 (1962), pp. 866–871. (Cited on pp. 2, 24, 26)

[29] H. Frahm, *Device for Damping Vibrations of Bodies*, U.S. Patent 989,958, Apr. 18, 1911. (Cited on p. 32)

[30] A. Giaralis and A. A. Taflanidis, *Optimal tuned mass-damper-inerter (TMDI) design for seismically excited MDOF structures with model uncertainties based on reliability criteria*, Struct. Control Health Monit., 25 (2017). (Cited on p. 38)

[31] R. M. Goodall and W. Kortüm, *Mechatronic developments for railway vehicles of the future*, Control Engrg. Practice, 10 (2002), pp. 887–898. (Cited on p. 36)

[32] E. A. Guillemin, *Introductory Circuit Theory*, Wiley, New York, 3rd ed., 1955. (Cited on p. 6)

[33] ———, *Synthesis of Passive Networks*, Wiley, New York, 1st ed., 1957. (Cited on pp. 2, 5, 14, 16)

[34] W. HÄHNLE, *Die Darstellung elektromechanischer Gebilde durch rein elektrische Schaltbilder*, in Wiss. Veröff. Siemens-Konz., vol. 11, Springer, Berlin, Heidelberg, 1932, pp. 1–23. (Cited on p. 29)

[35] A. HATCHER, *Algebraic Topology*, Cambridge University Press, Cambridge, UK, 2002. (Cited on p. 71)

[36] T. H. HUGHES, *Controllability of passive single-input single-output systems*, in Proceedings of the 2016 European Control Conference (ECC), 2016, pp. 1087–1092. (Cited on p. 27)

[37] ———, *Passivity and electric circuits: A behavioral approach*, IFAC-PapersOnLine, 50 (2017), pp. 15500–15505. (Cited on pp. 26, 27)

[38] ———, *A theory of passive linear systems with no assumptions*, Automatica, 86 (2017), pp. 87–97. (Cited on pp. 15, 26, 27)

[39] ———, *Why RLC realizations of certain impedances need many more energy storage elements than expected*, IEEE Trans. Automat. Control, 62 (2017), pp. 4333–4346. (Cited on pp. 2, 14, 19)

[40] T. H. HUGHES, A. MORELLI, AND M. C. SMITH, *Electrical network synthesis: A survey of recent work*, in Emerging Applications of Control and Systems Theory, Springer, Cham, 2018, pp. 281–293. (Cited on p. 19)

[41] T. H. HUGHES AND M. C. SMITH, *Algebraic criteria for circuit realisations*, in Mathematical System Theory – Festschrift in Honor of Uwe Helmke on the Occasion of His Sixtieth Birthday, 2013, pp. 211–228. (Cited on pp. 18, 25)

[42] ———, *On the minimality and uniqueness of the Bott-Duffin realization procedure*, IEEE Trans. Automat. Control, 59 (2014), pp. 1858–1873. (Cited on pp. 2, 14, 19)

[43] ———, *Controllability of linear passive network behaviors*, Systems Control Lett., 101 (2015), pp. 58–66. (Cited on pp. 15, 26)

[44] J. Z. JIANG, A. Z. MATAMOROS-SANCHEZ, R. M. GOODALL, AND M. C. SMITH, *Passive suspensions incorporating inerters for railway vehicles*, Vehicle Syst. Dyn., 50 (2012), pp. 263–276. (Cited on pp. 32, 36, 37)

[45] J. Z. JIANG, A. Z. MATAMOROS-SANCHEZ, A. ZOLOTAS, R. M. GOODALL, AND M. C. SMITH, *Passive suspensions for ride quality improvement of two-axle railway vehicles*, Proc. Inst. Mech. Eng. F, 229 (2015), pp. 315–329. (Cited on pp. 32, 36, 37)

[46] J. Z. JIANG, T. X. MEI, AND M. C. SMITH, *Curving performance for railway vehicles with advanced passive suspensions*, in Proceedings of the 23rd IAVSD, 2013. (Cited on pp. 36, 37, 38)

[47] J. Z. JIANG AND M. C. SMITH, *Regular positive-real functions and five-element network synthesis for electrical and mechanical networks*, IEEE Trans. Automat. Control, 56 (2011), pp. 1275–1290. (Cited on pp. 2, 19, 20, 21, 22, 23, 42, 45, 56, 62, 73, 85, 88, 89, 90, 95, 97, 100, 102, 103, 104, 107, 116, 117, 119, 120)

[48] ———, *On the theorem of Reichert*, Systems Control Lett., 61 (2012), pp. 1124–1131. (Cited on pp. 22, 23, 24, 78)

[49] ———, *Series-parallel six-element synthesis of biquadratic impedances*, IEEE Trans. Circuits Syst., 59 (2012), pp. 2543–2554. (Cited on pp. 19, 20, 61)

[50] J. Z. Jiang, M. C. Smith, and N. E. Houghton, *Experimental testing and modelling of a mechanical steering compensator*, in Proceedings of the 3rd International Symposium on Communications, Control and Signal Processing, Malta, 2008, pp. 249–254. (Cited on pp. 32, 35, 36)

[51] R. Kalman, *Old and new directions of research in system theory*, in Perspectives in Mathematical System Theory, Control, and Signal Processing, Lect. Notes Control Inf. Sci. 398, Springer, Berlin, Heidelberg, 2010, pp. 3–13. (Cited on pp. 2, 20, 25)

[52] ———, *Algebraic invariant theory in systems research (Lecture 1)*, in Proceedings of the 21st International Symposium on Mathematical Theory of Networks and Systems, Groningen, The Netherlands, 2014, pp. 1116–1117. (Cited on p. 20)

[53] ———, *Algebraic invariant theory in systems research (Lecture 2)*, in Proceedings of the 21st International Symposium on Mathematical Theory of Networks and Systems, Groningen, The Netherlands, 2014, p. 1118. (Cited on p. 20)

[54] A. E. Kennelly, *The equivalence of triangles and three-pointed stars in conducting networks*, Electrical World Engineer, 34 (1899), pp. 413–414. (Cited on p. 46)

[55] H. E. Koenig and W. A. Blackwell, *Electromechanical System Theory*, McGraw–Hill, New York, 1st ed., 1961. (Cited on p. 30)

[56] E. L. Ladenheim, *A Synthesis of Biquadratic Impedances*, Master's thesis, Polytechnic Institute of Brooklyn, 1948. (Cited on pp. 2, 19, 41, 42)

[57] I. F. Lazar, S. Neild, and D. Wagg, *Using an inerter-based device for structural vibration suppression*, Earthquake Engrg. Struct. Dyn. 43 (2014), pp. 1129–1147. (Cited on p. 38)

[58] D. J. N. Limebeer and R. S. Sharp, *Bicycles, motorcycles, and models*, IEEE Control Syst. Mag., 26 (2006), pp. 34–61. (Cited on p. 35)

[59] S. Lin, L. Oeding, and B. Sturmfels, *Electric Network Synthesis*, UC Berkeley, 2011, http://www.acritch.com/media/bass/electric-network-synthesis-v2.pdf. (Cited on pp. 20, 74, 75)

[60] L. Marian and A. Giaralis, *Optimal design of a novel tuned mass-damper-inerter (TMDI) passive vibration control configuration for stochastically support-excited structural systems*, Probab. Engrg. Mech., 38 (2014), pp. 156–164. (Cited on p. 38)

[61] R. W. Newcomb, *Linear Multiport Synthesis*, McGraw–Hill, New York, 1st ed., 1966. (Cited on p. 11)

[62] R. H. Pantell, *A new method of driving point impedance synthesis*, Proc. IRE (Correspondence), 42 (1954), p. 861. (Cited on pp. 2, 14)

[63] C. Papageorgiou, O. G. Lockwood, N. E. Houghton, and M. C. Smith, *Experimental testing and modelling of a passive mechanical steering compensator for high-performance motorcycles*, in Proceedings of the European Control Conference, Kos, Greece, 2007, pp. 3592–3599. (Cited on p. 36)

[64] C. Papageorgiou and M. C. Smith, *Laboratory experimental testing of inerters*, in Proceedings of the 44th IEEE Conference on Decision and Control, Seville, 2005, pp. 3351–3356. (Cited on p. 30)

[65] ———, *Positive real synthesis using matrix inequalities for mechanical networks: Application to vehicle suspension*, IEEE Trans. Control Syst. Technol., 14 (2006), pp. 423–435. (Cited on pp. 32, 34)

[66] J. W. Polderman and J. C. Willems, *Introduction to Mathematical Systems Theory*, Springer, New York, 1st ed., 1998. (Cited on p. 26)

[67] M. Reichert, *Die kanonisch und übertragerfrei realisierbaren Zweipolfunktionen zweiten Grades (Transformerless and canonic realisation of biquadratic immittance functions)*, Arch. Elektron. Übertrag., 23 (1969), pp. 201–208. (Cited on pp. 18, 21, 22, 23, 78, 119)

[68] F. M. Reza, *Synthesis without ideal transformers*, J. Appl. Phys., 25 (1954), pp. 807–808. (Cited on pp. 2, 14)

[69] P. I. Richards, *A special class of functions with positive real part in a half-plane*, Duke Math. J., 14 (1947), pp. 777–786. (Cited on p. 13)

[70] J. Riordan and C. E. Shannon, *The number of two-terminal series-parallel networks*, J. Math. Phys., 21 (1942), pp. 83–93. (Cited on pp. 6, 42, 79)

[71] K. H. Rosen, *Discrete Mathematics and Its Applications*, McGraw–Hill, New York, 7th ed., 2012. (Cited on pp. 72, 73)

[72] E. J. Routh, *Dynamics of a System of Rigid Bodies*, Macmillan, London, 7th ed., 1905. (Cited on p. 7)

[73] F. Scheibe and M. C. Smith, *Analytical solutions for optimal ride comfort and tyre grip for passive vehicle suspensions*, Vehicle Syst. Dyn., 47 (2009), pp. 1229–1252. (Cited on pp. 32, 34)

[74] S. Seshu and M. B. Reed, *Linear Graphs and Electrical Networks*, Addison–Wesley, Reading, MA, 1st ed., 1961. (Cited on pp. 6, 54)

[75] J. L. Shearer, A. T. Murphy, and H. H. Richardson, *Introduction to System Dynamics*, Addison–Wesley, Reading, MA, 1st ed., 1967. (Cited on p. 30)

[76] A. Siami, H. R. Karimi, A. Cigada, E. Zappa, and E. Sabbioni, *Parameter optimization of an inerter-based isolator for passive vibration control of Michelangelo's Rondanini Pietà*, Mech. Syst. Signal Process., 98 (2018), pp. 667–683. (Cited on pp. 32, 33)

[77] P. Slepian and L. Weinberg, *Synthesis applications of paramount and dominant matrices*, in Proceedings of the National Electronics Conference, vol. 14, Chicago, 1958, pp. 611–630. (Cited on p. 18)

[78] M. C. Smith, *Synthesis of mechanical networks: The inerter*, IEEE Trans. Automat. Control, 47 (2002), pp. 1648–1662. (Cited on pp. 19, 29, 30, 32)

[79] ———, *Force-Controlling Mechanical Device*, U.S. Patent 7,316,303, Jan. 8, 2008. (Cited on p. 30)

[80] ———, *Inerters*, in The Princeton Companion to Applied Mathematics, Princeton University Press, Princeton, NJ, 2015, pp. 604–609. (Cited on p. 33)

[81] ———, *Kalman's last decade: Passive network synthesis*, IEEE Control Syst. Mag., 37 (2017), pp. 175–177. (Cited on p. 20)

[82] M. C. Smith and S. J. Swift, *Design of passive vehicle suspensions for maximal least damping ratio*, Vehicle Syst. Dyn., 54 (2016), pp. 568–584. (Cited on pp. 34, 35)

[83] M. C. SMITH AND G. W. WALKER, *Performance limitations and constraints for active and passive suspensions: A mechanical multi-port approach*, Vehicle Syst. Dyn., 33 (2000), pp. 137–168. (Cited on p. 33)

[84] M. C. SMITH AND F. C. WANG, *Controller parameterization for disturbance response decoupling: Application to vehicle active suspension control*, IEEE Trans. Control Syst. Technol., 10 (2002), pp. 393–407. (Cited on p. 33)

[85] ———, *Performance benefits in passive vehicle suspensions employing inerters*, Vehicle Syst. Dyn., 42 (2004), pp. 235–257. (Cited on pp. 33, 34)

[86] S. J. SWIFT, M. C. SMITH, A. R. GLOVER, C. PAPAGEORGIOU, B. GARTNER, AND N. E. HOUGHTON, *Design and modelling of a fluid inerter*, Internat. J. Control, 86 (2013), pp. 2035–2051. (Cited on p. 30)

[87] M. E. VAN VALKENBURG, *Introduction to Modern Network Synthesis*, Wiley, New York, 1st ed., 1960. (Cited on pp. 5, 12)

[88] C. G. VASILIU, *Series-parallel six-element synthesis of the biquadratic impedances*, IEEE Trans. Circuit Theory, 17 (1970), pp. 115–121. (Cited on p. 79)

[89] ———, *Correction to "Series-parallel six-element synthesis of the biquadratic impedances,"* IEEE Trans. Circuit Theory, 18 (1971), p. 207. (Cited on p. 79)

[90] F. C. WANG, C. W. CHEN, M. K. LIAO, AND M. F. HONG, *Performance analyses of building suspension control with inerters*, in Proceedings of the 46th IEEE Conference on Decision and Control, New Orleans, 2007, pp. 3786–3791. (Cited on p. 38)

[91] F. C. WANG, M. F. HONG, AND C. W. CHEN, *Building suspensions with inerters*, Proc. Inst. Mech. Eng. C, 224 (2010), pp. 1605–1616. (Cited on pp. 32, 38)

[92] F. C. WANG, M. F. HONG, AND T. C. LIN, *Designing and testing a hydraulic inerter*, Proc. Inst. Mech. Eng. C, 225 (2011), pp. 66–72. (Cited on p. 30)

[93] F. C. WANG, M. R. HSIEH, AND H. J. CHEN, *Stability and performance analysis of a full-train system with inerters*, Vehicle Syst. Dyn., 50 (2012), pp. 545–571. (Cited on p. 36)

[94] F. C. WANG AND C. H. LEE, *Vibration suppression of milling machine tools by the inerter*, in Proceedings of the 23rd International Symposium on Mathematical Theory of Networks and Systems, Hong Kong, 2018, pp. 626–630. (Cited on p. 32)

[95] F. C. WANG AND M. K. LIAO, *The lateral stability of train suspension systems employing inerters*, Vehicle Syst. Dyn., 48 (2010), pp. 619–643. (Cited on pp. 32, 36, 37)

[96] F. C. WANG, M. K. LIAO, B. H. LIAO, W. J. SU, AND H. A. CHAN, *The performance improvements of train suspension systems with mechanical networks employing inerters*, Vehicle Syst. Dyn., 47 (2009), pp. 805–830. (Cited on pp. 36, 37)

[97] H. WHITNEY, *2-Isomorphic graphs*, Amer. J. Math., 55 (1933), pp. 245–254. (Cited on pp. 6, 42)

[98] ———, *On the classification of graphs*, Amer. J. Math., 55 (1933), pp. 236–244. (Cited on pp. 6, 42)

[99] J. C. WILLEMS, *The behavioral approach to open and interconnected systems*, IEEE Control Syst. Mag., 27 (2007), pp. 46–99. (Cited on pp. 19, 26)

[100] ———, *Dissipative dynamical systems*, Eur. J. Control, 13 (2007), pp. 134–151. (Cited on pp. 19, 26)

[101] K. YAMAMOTO AND M. C. SMITH, *Bounded disturbance amplification for mass chains with passive interconnection*, IEEE Trans. Automat. Control, 61 (2016), pp. 1565–1574. (Cited on pp. 32, 38)

[102] D. C. YOULA, *Theory and Synthesis of Linear Passive Time-Invariant Networks*, Cambridge University Press, Cambridge, UK, 1st ed., 2015. (Cited on pp. 5, 17, 25)

[103] D. C. YOULA AND P. TISSI, *N-port synthesis via reactance extraction, Part* I, IEEE International Convention Record, 7 (1966), pp. 183–205. (Cited on pp. 17, 18)

[104] S. Y. ZHANG, J. Z. JIANG, AND S. A. NEILD, *Passive vibration control: A structure-immittance approach*, Proc. R. Soc. A, 473 (2017), 20170011. (Cited on pp. 34, 35)

[105] S. Y. ZHANG, J. Z. JIANG, AND M. C. SMITH, *A new proof of Reichert's theorem*, in Proceedings of the 55th IEEE Conference on Decision and Control, Las Vegas, 2016, pp. 2615–2619. (Cited on p. 23)

[106] O. J. ZOBEL, *Theory and design of uniform and composite electric wave-filters*, Bell Syst. Tech. J., 2 (1923), pp. 1–46. (Cited on pp. 46, 67)

Index